Coastal Atlantic

Sea Creatures:

A Natural History

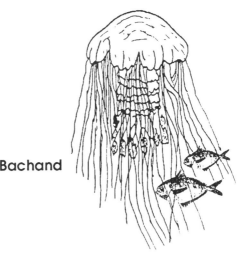

Robert G. Bachand

The Maritime Center
at Norwalk
10 North Water Street
Norwalk, CT 06854

Sea Sports Publications

Norwalk, CT

Estero, FL

Library of Congerss Cataloging-in-Publication Data
Bachand, Robert G.
Coastal Atlantic Sea Creatures: A natural history /
Robert G. Bachand
   p. cm.
Includes bibliographical references and index.
1. Coastal fauna--Atlantic Coast (U.S.) I. Title.
QL 155.B24 1994   94-34046
591.974--dc20   CIP

ISBN 0-9616399-5-4

To Kathy

Acknowledgements

I wish to thank to following individuals who reviewed parts or all of my manuscript for accuracy and made suggestions for the inclusion of further information. These individuals are not responsible for errors I may have made, but they must be given credit for the quality of the text.

Allen Bejda, National Marine Fisheries Service, NJ.
Walter Blogoslawski, National Marine Fisheries Service, CT.
Robert H. Brewer, Trinity College, Hartford, CT.
Gerard Capriulo, SUNY Purchase, NY.
David Conover, SUNY Stony Brook, NY.
Grant Gilmore, Harbor Branch Oceanographic Institute, FL.
Roger Hanlon, Marine Biomedical Institute, TX.
Eric Lazo-Wasen, Yale University, CT.
Roderick MacLeod, CT DEP, Fisheries Division, CT.
Linda Mercer, NC Division Marine Fisheries, NC.
Edward Monahan, CT Sea Grant, Avery Point, CT.
Carter Newell, Greater Eastern Mussel Farms, ME.
William Palmer, NC Museum Natural History, NC.
Mike Shirley, Rookery Bay NERR, FL.
Joe Schnerlein, Brien McMahon High School, Norwalk, CT.
Jack Schneider, The Maritime Center at Norwalk, CT.
Eric Smith, CT DEP, Fisheries Division, CT.
James Stone, Mystic Marinelife Aquarium, Mystic, CT.
Robert Trifone, Brien McMahon High School, Norwalk, CT.
Barbara Welsh, University of Connecticut, Avery Point, CT.
Howard M. Weiss, Project Oceanology, Avery Point, CT.
Jim Widman, National Marine Fisheries Service, CT.
and
Joe Schnerlein's students at Brien McMahon High School who made comments on the text's readability and its illustrations.

In pursuing my passion for the marine environment, I must particularly thank my wife Kathy. Over these many years, she has put up with my frequent weekend SCUBA dives, a family room occupied by small aquariums and jars of live zooplankton stored in our refrigerator.

All photographs and illustrations by the author unless otherwise indicated.

TABLE OF CONTENTS

Foreword

As I began to read Robert Bachand's attractive book, *Coastal Atlantic Sea Creatures: A Natural History*, a thought immediately came to mind. Here is a book that I wish I had at hand when as a teenager I spent time at the beach, and poking about in the salt marshes near my home. But as I read further, and gave some more thought to it, I quickly recognized that, from my vantage as a by-no-means-young marine scientist whose technical specialty is other than zoology, this book is both informative and appealing.

Part of the book's appeal for me is intrinsic to the subject covered, and clearly in part arises out of the author's love and enthusiasm for this material, which he effectively conveys with an economy of words. Bachand's goal in writing this book was to provide under one cover the natural history of a wide selection of our coastal creatures, in a style appropriate for amateur naturalists, teachers of science in our schools and members of the general public. I am confident that he has, with this comparatively slim volume, achieved his aim.

By dividing the book into three sections, one on animals found in the mud flats and salt marshes, a second on the creatures of sandy substrats, and the third on the inhabitants of the rocky environments, Bachand has added to the utility, and ease-of-use, of what was already a handy book.

I like this book. I've learned from it, and I expect that others who share my enthusiasm and concerns for the various aspects of our marine environment, will do likewise. Anyone heading off to spend a few hours, or a few days, along our coast would be well advised to find room in their pack for a good field guide, and for a copy of Bachand's delightful natural history volume.

Edward C. Monahan, PhD., D.Sc.
Professor of Marine Sciences, University of Connecticut
Director, Connecticut Sea Grant College Program
March, 1994

Preface

*S*ome years ago, prior to the opening of the Maritime Center at Norwalk, I helped edit and co-write copy for the aquarium's information rails. During this process, I became aware that, in the popular literature, there was a lack of detailed information concerning the natural history of most marine animals. A wealth of such material exists in scientific journals, but very few authors have translated it for the amateur naturalist or the general public. *Coastal Atlantic Sea Creatures: A Natural History* covers a selected number of the fascinating and varied marine animals that inhabit our shoreline. It examines their role in the coastal ecosystem, adaptations to the stresses of their environment, species interrelationships and reproductive strategies.

The book is divided into the three major shoreline and nearshore environments: Mud (including the salt marsh), Sand and Rock. Many of the species don't fit neatly in any single environment, and in those cases, a specific environment was selected for a phase in the animal's life cycle. The choice of animals included in the book was a difficult one. Selection was based on geographic distribution, (how common), commercial importance, unusual behavior, mode of reproduction and/or perceived reader interest. There are obviously many other animals that could have been covered. But it is the hope of this marine naturalist that the sampling of *Coastal Atlantic Sea Creatures* will stimulate the reader to search out information on other inhabitants of our shoreline.

Robert (Bob) G. Bachand
Norwalk, CT
September, 1994

CREATURES OF THE TIDAL MUD FLAT AND SALT MARSH

1

*W*hat could possibly live in those foul-smelling, sulphurous muds? Oxygen barely penetrates the top 3/8 inches (10 mm) of the sediments and anything deeper reeks of rotten eggs! At low tide, the animals living on the exposed surfaces are subject to the blistering sun. When it rains, they are inundated by fresh water. During the winter, these same creatures can be stressed by sub-zero temperatures or they can be carried off or torn to shreds by rafts of ice.

Despite the rigors, the muds of the tidal flat and marsh are host to a variety of marine organisms that include algae, bacteria, protozoans, worms, snails, clams, crustaceans and fishes. Similar to the sand and rock environments, some of the residents of the tidal mud flat and salt marsh are restricted to a specific site while others are able to move in and out with the seasons. There are also those that have the ability to thrive in various environments. Some of the marine animals swim above the bottom or live on its surface (epifauna). Others live within the sediment (infauna).

Salt marshes and tidal mud flats form in protected areas behind barrier beaches or other natural structures. There, in quiet

1

waters, fine sediments have time to settle out and marsh plants can establish themselves. The muds at these sites tend to be composed mostly of clay and silt, with large accumulations of dead or dying plants and animals (detritus). For many of the animals living in or on the mud, detritus serves as a food source. These animals in turn serve as prey for fish, crabs and a variety of birds and small mammals.

Tidal mud flats are a familiar feature of the New England shoreline. In Maine, they represent about 48 percent of the state's intertidal habitats. The shoreward edges of some tidal mud flats are bounded by salt marshes and their seaward edges are bounded by sand flats, tidal channels or eelgrass beds.

Along the East Coast, salt marshes are found from Newfoundland to Florida. In the more tropical areas of Florida, however, these habitats are replaced by mangroves. The low or regularly flooded salt marshes support wide expanses of salt marsh cordgrass, *Spartina alterniflora*. On the north and mid-Atlantic coast, the high or infrequently flooded marshes are covered by saltmeadow hay, *Spartina patens*, but in Georgia a stunted form of cordgrass replaces saltmeadow hay.

Spartinas constitute an important part of the coastal food chain. Yet, with the exception of the purple marsh crab and some insects, very few creatures feed directly on them. Before these grasses can be consumed by most animals, they must first be broken down into detritus, a process that involves bacteria, protozoa and fungi.

Tidal mud flat fringed by *Spartina alterniflora*, Shea Island, CT.

2

Saltmeadow hay detritus tends to remain in the high marsh where it serves as food for residents such as the saltmarsh snail. In the low marsh, most of the detrital cordgrass is swept out by the tides and is deposited on the mud flats or is carried to open water.

It has been estimated that one acre of salt marsh produces 10 tons of organic matter annually; the average annual yield of one acre of farmland is only about one and a half tons. As will be seen, however, the marsh is not just a food source, it is also a important nursery for many of our coastal invertebrates and fishes. ■

**Salt marsh cordgrass,
Spartina alterniflora.**

Saltmeadow hay, *Spartina patens*, Stratford, CT.

3

The invertebrates

Salt-marsh snail, (=Eastern melampus, AFS, 1988)
Melampus bidentatus
Habitat: Upper reaches of salt marshes.

> Other common names: Coffee bean snail.
> Phylum: Mollusca. Class: Gastropoda.
> Order: Archaeopulmonata. Family: Melampodidae.
> Geographic range: New Brunswick to Texas.
> Size: Maximum size .39 to .47 inches (10-12 mm).
> Reproductive season: Lunar spring tides.
>   North of Cape Cod, MA, June.
>   Falmouth, MA, late May to early July.
>   Cape Hatteras, NC, August.
>   Fort Macon, NC, August-September.
> Egg production: Deposited in a single continuous
>   strand composed of an average of 850 eggs.
> Life span: 3 to 4 years.

*I*t happens twice a month. The sun, moon and earth align themselves, and produce higher than normal high tides and lower than normal low tides (spring tides). During the full moon and the new moon, rising waters of a spring tide invade the upper reaches of the marsh and flood the realm of the salt-marsh snail. Many of this unusual mollusk's activities are synchronized to the rhythms of spring tides.

The salt-marsh snail belongs to a group that includes the most primitive living pulmonate (lung) snails; it breathes air as do

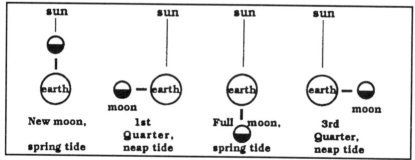

Lunar tidal cycle: Many of the coastal creatures time their reproductive activities to a lunar tide. When the moon and sun are in line with each other, the resulting gravitational forces combine to produce tides that are higher than normal and lower than normal; these are known as spring tides. When the moon and sun are at right angles to each other, the gravitational forces counteract each other. The result is a smaller difference (tidal range) between high and low tides; these are known as neap tides.

4

Salt-marsh snail, length 3/8 inch. Sherwood Island State Park, CT.

land snails. A pulmonate snail's mantle[1] cavity has a tiny opening known as the pneumostome. Gills are absent in this mollusk. The roof of its mantle cavity, which is highly vascular, forms a respiratory surface. Air is moved in and out through the pneumostome by alternately arching and flattening the mantle floor.

The salt-marsh snail is well adapted to life in the highest tidal zone of the marsh. To escape the rising waters, it often climbs the stalks of saltmeadow hay. But if trapped below water, the animal can remain submerged for as long as 14 days without drowning. It breathes through its skin while still submerged.

The snail is most active at night, taking advantage of lower air temperatures and higher relative humidity. Though it apparently can withstand temperatures of up to 112.5°F (30.5°C), it generally seeks refuge from the heat of the day by hiding under clumps of saltmeadow hay or any other cool and moist site. During the winter, the snail is said to burrow into the marsh soil or enters empty fiddler crab burrows. It can, however, tolerate temperatures of 14°F (-10°C) by producing an antifreeze and other biochemicals that direct ice formation between cells (extracellular) rather than inside cell walls (intracellular). This ability allows the snail to survive with nearly 75 percent of its tissue fluids frozen!

It is during the salt-marsh snail's reproductive season that the creature exhibits yet another remarkable adaptation. In synchrony

[1] The mantle is a fleshy fold of body wall that forms a space (mantle cavity). The mantle secretes the shell.

with a new or full moon, the snails gather in large groups to mate. Each is simultaneously both male and female (functional hermaphrodite); individuals assume the role of one sex and then the other sometime during the breeding cycle. Though mating pairs are capable of fertilizing each other at the same time, only one is usually impregnated.

Egg laying is believed to occur one to three days after mating. The snail produces a single continuous strand of some 600 to 1200 eggs that it deposits on the marsh surface or on the stems and leaves of the saltmeadow hay. Within a short time, the action of the tide covers the strand with organic debris and sediments, thereby protecting the eggs from drying out.

The eggs hatch during the next spring tide, some 13 days after being laid. Following several successive floodings, the eggs rupture and tiny larvae escape in the receding tide. Weakly propelled by a spinning fringe or veil, the larval snails (veligers, see Free-swimming larval veliger) drift, feed and grow in the inshore plankton (also see Eastern mud snail, page 7). They remain in the plankton[2] for two weeks to six weeks. Those that survive and find their way back to a marsh, settle in an area already occupied by adults. Within a few hours of becoming bottom-dwelling creatures, the snails begin to consume a variety of algae and plant material. They process the materials and, as an important part of the marsh ecosystem, many of them become food for a variety of fishes, crabs and sea birds. ■

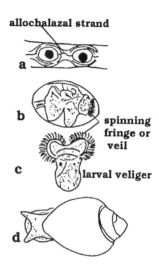

allochalazal strand

a

b

spinning fringe or veil

c

larval veliger

d

a: Egg strand of the salt-marsh snail.
The eggs are protected by two gelatinous layers. A fine thread-like structure (allochalazal strand) connects one egg to the next; it apparently functions mechanically in helping larvae break away from their egg shell during hatching.

b: Larval veliger developing within the egg.

c: Free-swimming larval veliger. Its shell is approximately 125 microns in diameter. The veliger remains in the near-shore plankton until the next spring tide.

d: Young salt-marsh snail spat. Approximately 4 to 5 weeks after settling.

After Russell-Hunter, 1972.

Number of salt-marsh snails per square meter.
Canary Creek, DE: 693/m² +-340
Dras Creek, NJ: 445/m² +-184
Poropotank River, VA: 144/m²

[2] Plankton are tiny, often mocroscopic organisms that drift in the currents.
Zooplankton = animal plankton.
Phytoplankton = plant plankton.

Eastern mud snail,
Ilyanassa (=Nassarius) obsoleta

Habitat: Intertidal mudflats to just below the low tide line.

> Other common names: Mud snail.
> Phylum: Mollusca. Class: Gastropoda.
> Order: Neogastropoda. Family: Nassariidae.
> Geographic range: Gulf of Saint Lawrence to Florida.
> Salinity: Inhabits sites exceeding 15 ppt.
> Egg capsule: About 2.7 mm in height.
> Egg production: Each capsule contains 40 to 150+ eggs.
> Larval stage: 250 microns in length at time of emergence.
>   Distinguishing characteristic - red pigmented border on velum.
> Life span: 3 years - a few may live up to 8 years.

*E*xtending her long powerful foot, the mud snail makes her way across the mud flat to solid ground. Once there, the female carefully molds and cements straw-colored eggs capsules to a rock or algae; each structure is filled with some 100 eggs. Sheltered within the confines of their protective cases, the developing mud snails begin to show signs of life on the second day. Over the next few days, the creatures increase their activity. They spin around and bump into each other as they prepare to become temporary members of the plankton community. By the eighth day, the larvae release a chemical that softens the plug at the top of the capsule. Then, within a short time, they escape into a realm where relatively few survive to maturity.

Prior to hatching, the larval snails develop a veil-like membrane (velum) equipped with opposing bands of hair-like cilia at the outer edges. The veligers, as the larvae are called, have two eyes, a slit-shaped mouth and a shield-shaped foot. The cilia help propel the veligers through the water. The creatures can hover in place, dart along at an amazing pace or when the velum is retracted somewhat, they sink toward the bottom. Their ability to swim, however, is no match for the currents. They thus drift along with the rest of the plankton population.

The veligers are tiny eating machines. As they swirl through the water, they capture food by concentrating it between their two bands of cilia. A small groove between the bands directs the particles toward the mouth. The veligers spend several weeks in the plank-

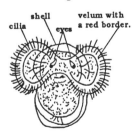

Newly hatched mud snail veliger.
After Scheltema, 1962.

7

ton. Those that find an adequate supply of microscopic algae and do not fall victim to predators or the elements, begin to search for a place to settle.

During this time, the larvae are known as pediveligers. They have the ability to alternately creep on the bottom, swim and delay metamorphosis into an exclusively bottom-dwelling creature for as long as 20 days. The behavior gives the larval snails a better chance of finding a suitable home. Once they have settled to the bottom, they cast off their velum, a process that takes about 20 to 30 minutes. During a favorable year, the newly metamorphosed snails have been observed in concentrations of as many as 23,000 individuals per square meter! As with most species of snail

Eastern mud snail's egg capsule.

An individual snail's egg capsules are identical to each other, but they are different from those produced by other Eastern mud snails.

During the spring, egg capsules can be found intertidally attached to seaweeds or solid surfaces. Activity within the capsules can be seen with the aid of a low-power microscope.

that produce vast numbers of offspring, however, a very small percentage survive to their first birthday. For those that do make it to adulthood, their life span can reach three or more years.

An 1873 survey of the Cape Cod region found the adult mud snail "...dominant on sand and mud flats, pilings, sea walls, salt marshes, and eel grass beds, and common on protected rocks, cobble beaches, and pilings" (Brenchley, 1983). Two decades later, the mud snail had been displaced by the invading common periwinkle and then became confined primarily to soft mud/sand flats (see Common periwinkle, page 86).

Despite the loss of much of its former habitat, the mollusk has continued to flourish. Sometimes described as a non-selective biological vacuum cleaner, the animal's continued survival as one of the most common snails of the northeast Atlantic coast is at least partially due to its ability to feed on almost anything. The mud snail is primarily a deposit feeder, consuming microscopic algae found on the surface sediments of tidal flats, but it is also a scavenger that is quickly attracted to dead fish, crabs or mollusks.

How the mud snail finds carrion and escapes predators is another interesting facet of its life. The mud snail easily detects a dead creature by chemical cue. Moving up-current toward the source, it waves its siphon back and forth. Once it has approached the carrion to within an inch or less, the mud snail extends its tubelike proboscis. Using a file-like radula that is housed in the proboscis, the snail begins to scrape away bits of the dead animal's tissues.

Even snails that are buried under the mud at low tide, emerge when they become aware of carrion. In a short time, hoards of snails are feeding on the fallen animal. Using chemical cues, mud snails can also detect the crushing and shredding of one of their own by a predator such as the green crab. The response is immediate. The alarmed creatures scatter in all directions, travelling of course, at a snail's pace. Within few minutes a circular area of about four inches has been cleared of the mollusks. Some use their long powerful foot to bury themselves. They disappear below the mud surface in less than five seconds and leave little evidence of their presence. Others, however, continue to move away for up to two hours. The alarm substance that is apparently present in all of the mud snail's tissues and blood, can persist in the sediments for more than 16 hours. For as long as it remains, the alarm substance continues to affect the creature's behavior.

The approaching winter initiates yet another behavior in the mud snail. Gathering in large groups, the creatures make their way to the low tide zone where they number in the hundreds to thousands. Covered by mud, the snails remain there over the winter and re-emerge in the spring. Some snails, however, do not make the seasonal migration and remain year-round in the high tidal zone.

Eastern mud snail, *Ilyanassa obsoleta. Newport, RI.*

9

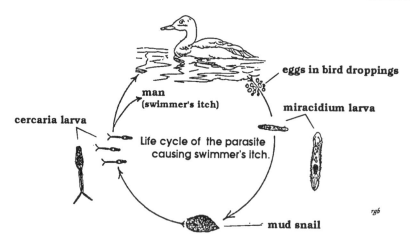

eggs in bird droppings

man
(swimmer's itch)

miracidium larva

cercaria larva

Life cycle of the parasite
causing swimmer's itch.

mud snail

These tend to be infected by a troublesome parasite (avian blood fluke) that can cause a skin rash in humans. The rash is variously known as swimmer's or clam-digger's itch, and schistosome dermatitis.

The culprit in swimmer's itch is the invasive larval stage (cercaria) of a parasitic trematode worm. The worm's natural hosts include the black duck, lesser scaup duck, red-breasted merganser and possibly other sea birds. The mud snail is merely an intermediate host in its life cycle and people are its accidental victim. As the spring waters rise to about 50°F (10°C), the cercaria emerges from an infected snail and penetrates a bird's skin. It makes it way from the skin to the blood stream and migrates to the veins surrounding the intestines. By then, the larvae that have metamorphosed into adult worms, mate and deposit eggs within the intestinal walls. The eggs then find their way into the environment inside the bird's droppings. When the droppings fall into the sea, the eggs hatch and produce larvae known as miracidium. These larvae, which only infect the mud snail, produce a series of saclike structures (sporocysts) within their host's digestive and sex glands. The sporocysts in turn produce cercaria larvae, completing the cycle.

When humans come into contact with the 1/50 inch cercaria, the parasite is quickly killed in the skin by the body's natural defenses. Over twenty species of the parasites are known to cause swimmer's itch, in both fresh and salt water. Three that are found in tropical and subtropical parts of the world, produce serious disease in humans. Luckily, the local parasite known as *Austrobilharzia variglandis*, causes only a skin rash that disappears within a short time. The severity of the rash seems to depend on previous exposure. Repeated contact leads to more severe cases.

Bay scallop,
Argopecten (=Pecten,=Aequipecten) irradians
Habitat: Eelgrass beds and sandy/mud bottoms.

Other common names: Blue-eyed scallop.
Phylum: Mollusca. Class: Bivalvia.
Order: Ostreoida. Family: Pectinidae.
Geographic range of the three subspecies of *Argopecten*:
A. i. *irradians*: Cape Cod, MA, to New Jersey.
A. i. *concentricus*: New Jersey to Chandeleur Islands,
                Gulf of Mexico.
A. i. *amplicostatus*: Galveston, TX to Laguna Madre, TX.
Depth range: Low tide line to about 33 feet (10 m) of water.
Salinity: Minimum salinity tolerance approximately 14 ppt.
Reproduction - sexes: Hermaphroditic.
Reproductive season:
        Long Island Sound: Mainly in June and July.
        North Carolina & Florida: August to December.
Eggs: Unfertilized eggs average 60 millimicrons in diameter.
        Pink to red in color.
Egg development: Fertilized egg to settlement about 14 days.
Life span:
        South of Maryland, average 12 to 24 months.
        North of Maryland, average is 20 to 26 months.
        North of Connecticut, to 30 months.

*T*he shallows near the low tide line can be an ill-chosen habitat for a bay scallop. As the water drops toward its lowest point, herring gulls prowl the surface and snatch up any scallop within their reach. Ascending high over the beach, the birds release their catch and send it hurling toward the rocks below. If the shell doesn't break, the gulls pick it up and rise skyward again. They may repeat the behavior several more times before succeeding. Then the spoils of victory are consumed. In deeper water, there is safety from marauding birds. But even there predators such as the green crab (*Carcinus maenas*), oyster drill (*Urosalpinx cinerea*), sea stars (*Asterias* spp.) and humans seek out this delectable bivalve mollusk.

The bay scallop is nonetheless well equipped for protecting itself from most of its enemies; it is alerted to danger using its chemical, tactile and/or visual senses. Researchers have shown that the creature can detect crude chemical extracts of sea star. The long tentacles that fringe the scallop's mantle are sensitive to a predator's touch. The eyes that line the edges of the mantle folds are known to alert the animal to sudden changes in light intensity. It is these beautiful and arresting, steel-blue eyes that give the creature its name, blue-eyed scallop.

11

Each of the scallop's eyes has a cornea, lens, double-layered retina and optic nerve. The eyes cannot form an image but their lenses act as condensers, concentrating light rays. One layer of the retina contains receptors that respond to increases in light intensity and the other reacts to a decrease. When a shadow is cast across a single eye, the animal generally shows no response, but one that moves from eye to eye can elicit a reaction. If a predatory snail or some other threatening creature touches the scallop's tentacles, the mollusk may initially close its valves. When sufficiently alarmed, it escapes by swimming off the bottom. The scallop may, however, to be able to discriminate between certain predators and non-predators. The ability no doubt helps it save energy for real emergencies.

The scallop's swimming behavior is usually caused by a predator's attack. Slamming its valves shut by the rapid contraction of its adductor muscle, the creature expels jets of trapped water from both sides in the back of the shell. The force propels it off the bottom. Using the curtain-like velum that lines the edges of both shells, the animal can direct water alternately from the back and the front of the shell. Though the action sends it in a typical zigzag-like trajectory, the scallop is nevertheless believed to have some control over its movement. When water is released from only one side of the shell, the creature moves sideways. It generally covers just a short distance, and it is not known to migrate to any extent from one site to another. In a survey of the Niantic Bay, CT, estuary, bay scallops moved less than three feet over a period of six days. The mollusc swims only as long as it claps its shells together. When it stops, it slips sideways, back and forth, much like a coin sinking to the bottom. For most of its adult life, however, the creature lies on the bottom with its right shell down and its valves opened approximately 20 degrees. In this position, it feeds.

water jets

direction of movement

The scallop is a filter feeder. It pumps water though its mantle cavity, from the front toward the back of the shell; a large scallop can reportedly move water at the rate of up to 6.7 gallons (25.4 L) per hour. As the current flows across the gills, food particles are trapped in mucous secreted by the gills. The materials are then moved toward the mouth by beating, hairlike cilia.

The mollusk consumes planktonic algae, benthic (bottom-dwelling) diatoms, bacteria and certain detritus. An adequate supply of suitable food and water temperature greatly affects reproduction in the scallop. During the mid-1980s, an unusual bloom of a minute planktonic alga occurred in the bays of Rhode Island, New York and New Jersey. The alga, a chrysophyte (2-3 microns in diameter), proved devastating to the scallop. The previously undescribed organism grew in such large numbers that the region's waters turned dark brown. Known as the brown tide, it blocked sunlight killing off much of the eelgrass in Peconic and Gardiners Bay, NY. The scallops were unable to digest the alga and many starved. The bloom also coincided with the area's reproductive period, resulting in the delay or failure to spawn; the larvae of those that did spawn often died during their short planktonic stage.

Along the coast of southern New England, New York and New Jersey, spawning occurs mainly during June and July, as the water temperature increases. In North Carolina and Florida, scallops reproduce from August to December as the waters cool. The scallop is hermaphroditic. Though an individual may release both reproductive products within a short time of each other, self fertilization is thought to be rare.

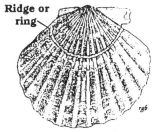

Ridge or ring

The ridge or ring in the surface of the scallop is due to slow growth during the winter of its first year of life.

Fertilization takes place in the water column or on the sea floor. Initially, the scallop's larva (trochophore) resembles the larva of an annelid worm (see Sandworm, page 66). The creature's structure is partially surrounded by short, hairlike cilia with a tuft of some six long cilia extending out from its widest end. Within 24 to 48 hours, the creature changes into a veliger larva. At this point it has two shells and a veil-like ring of short cilia that it uses to propel itself and gather food (see Eastern mud snail, page 6). The veliger remains in the plankton for some 10 days. As it drifts and feeds, it grows a muscular foot that allows it to settle on a surface, crawl about and swim off if the site is not suitable. This stage is known as pediveliger.

The young scallop settles on a variety of substrates including rocks, oyster shells, ropes, seaweeds and filamentous algae. Apparently, however, it prefers eelgrass, *Zostera marina*. Once settled, a special gland in the creature's foot produces a fine thread, byssus, by which it can attach itself. The scallop then climbs up toward the top of the eelgrass by extending its foot, grasping the surface and pulling itself up in a jerky motion. The upper eelgrass canopy helps the juvenile evade most benthic predators and keeps it free of silt that might otherwise smother it. The eelgrass community also benefits the scallop by slowing the current; filtering of algae from the water by the mollusk is most efficient in a slow current. When the growing

13

scallop has reached 0.8 to 1.2 inches (20 to 30 mm) in length, it drops to the bottom and adopts the life style of its parents. It spawns for the first time at the age of one, and it often does not live to see its second birthday. ■

Two-gilled bloodworm, *Glycera dibranchiata*

Habitat: Prefers soft muds rich in organic matter.

Other common names: Blood, bloodworm, beak-thrower, beakworm.
Phylum: Annelida. Class: Polychaeta.
Order: Phyllodocida. Family: Glyceridae.
Geographical range: Gulf of St. Lawrence to Florida. Gulf of Mexico.
Depth range: Intertidal near low tide line to 1322 feet (403m).
Color: Purplish-pink to red.
Size: Maximum 14.5 inches (370 mm).
Age at spawning: Approximately three years.
Sexual form: Epitoke.
Reproductive season: Nova Scotia and Maine - spring.
                    Maryland - November and possibly late spring.
Life span: Maximum 5 years; die after spawning.

*B*y any standard, the bloodworm is a formidable predator. It can detect vibrations in the soil produced by an approaching prey. Its four jaws are strengthened with copper, and each is coursed by a canal that ends in numerous pores. Through these pores it injects its neurotoxin, a poison that is especially effective on small crustaceans. The worm apparently feeds on dead and decomposing salt marsh plants and algae (detritus), but it has also been shown to actively prey on sandworms and certain amphipods (crustaceans). It is sometimes called the beak-thrower, a name that reflects the creature's ability to evert the front part of its digestive tract (proboscis) with its fang-like jaws set to strike. In the process, it can inflict a painful bite on an unsuspecting fisherman. Described as feeling like a bee sting, the wound can cause severe inflammation. Though some authors have questioned whether its diet consists mainly of detritus or of prey, there is no doubt that the bloodworm is well equipped to function as a predator.

The bloodworm tends to inhabit soft muds rich in organic matter but is also found in sand/mud bottoms and occasionally in fine sand. It spends most of its time hidden beneath the sediments, and it was generally assumed that it leaves the bottom only to reproduce. But a chance observation on the night of January 12, 1977, helped change that assumption. While breaking ice around aquaculture rafts on Damariscotta Cove, ME, a worker noticed that sandworms were swimming in the water/ice slurry. Further observations by David Dean, a researcher from the University of Maine, revealed that bloodworms were also in the water column. Neither of the species was spawning. After emerging from the sediments, bloodworms were seen being passively carried in the ebbing tide. The behavior, referred to as non-reproductive swimming, is probably used for migrating from site to site. It is during spawning, however, that the blood-worm is best known for its ability as a swimmer.

In preparation for spawning, the bloodworm is transformed into its free-swim-ming, sexual form (epitoke, see Sandworm, page 65). Its gut and musculature shrink dramatically, and its hair-like setae and the fleshy appendages (parapodia) along its sides lengthen. Shortly thereafter, on a daytime high tide, the worm rises from the bottom

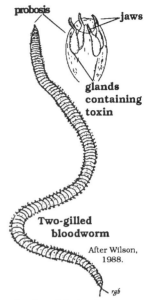

Two-gilled bloodworm

After Wilson, 1988.

Probosis with its four cop-per-hardened, black-colored jaws. Each jaw is hollow, al-lowing injection of toxin into a prey.

and begins its nuptial rite. During non-reproductive swimming, the worm propels itself in an awkward figure-eight motion. The spawner, however, moves in a graceful, undulating pattern.

Spawning reportedly occurs in shallow water, over a one to three-day period. The male swims slowly near the surface and releases two long streams of sperm; it then returns to the bottom. In contrast, the female swims rapidly near the surface. Within a short time, the posterior one-third of her body ruptures and up to 10 million eggs are released. The spent female is then barely able to swim, and she gradually sinks to the bottom. Both sexes die shortly after spawning. The remains of reproductive couples, consisting of the outer body wall and the everted proboscis with its intact jaws, are sometimes found intertidally. In this condition, they are known as ghost worms. But the dead worms do not go to waste. They become an opportune feast for striped bass and sand shrimp.

One of the most conspicuous predators of the live bloodworm is the black-bellied plover. Along the coast of Nova Scotia, the seabird has been observed displaying what is described as "foot-trembling" behavior. Apparently mistaking the vibration for an approaching prey, the worm moves to the surface and is quickly snatched-up by

15

the clever bird.

Bloodworms and sandworms are an important Atlantic coastal resource. They represent the fourth largest fishery in Maine, surpassed only by lobsters, clams and finfish. Some 90 percent of bloodworms are harvested in Maine. Nova Scotia accounts for 7 percent, and Massachusetts yields approximately 2 percent. ∎

Grass shrimp, *Palaemonetes* spp.

Habitat: Salt marshes, tidal creeks, eelgrass, seaweed beds, pilings and sandy shallows.

Marsh grass shrimp, *Palaemonetes vulgaris* (*P. vulgaris*)
Daggerblade grass shrimp, *Palaemonetes pugio* (*P. pugio*)
Brackish grass shrimp, *Palaemonetes intermedius* (*P. intermedius*)
Phylum: Arthropoda. Subphylum: Crustacea. Class: Malacostraca.
Order: Decapoda. Family: Palaemonidae.
Other common names: Glass shrimp, common shore shrimp,
                          prawn, hardback.
Geographic range:
         P. vulgaris - Southern Gulf of St. Lawrence to Yucatan.
         P. pugio - Nova Scotia to Texas.
         P. intermedius - Vineyard Sound, MA, to Texas.
Salinty tolerance: *P. vulgaris* more tolerant of high salinity
                 than *P. pugio.*
Egg production: *P. pugio* produces about 920 eggs/season.
Life span: About one year for *P. pugio* and *P. vulgaris.*

*T*railing its long, slender antennae, the nearly transparent grass shrimp moves effortlessly through the water. With its claws and legs extended, its tail (abdomen) arched slightly and its tailfan spread out for stability, the crustacean hovers in place or slowly propels itself forward using its swimmerets (pleopods). When a predator approaches, the shrimp snaps its tail and thrusts itself backward through the water. If sufficiently threatened, it can launch itself out of the water for a short distance. Should the creature become stranded along the edges of a tidal creek, it is generally able to return to the water by snapping its tail. In captivity, the grass shrimp often jumps out of the aquarium to escape any fish with a mouth large enough to swallow it. In nature, it can be assumed that all three species of grass shrimp residing along the Atlantic coast are food for the many estuarine fishes.

The three species of grass shrimp, marsh grass shrimp, daggerblade grass shrimp and brackish grass shrimp, share similar

16

habitats and geographical range. Their different tolerances to salinity, however, sometimes separate them. Closely related, they can be distinguished from each other by their rostrum or anterior spine (see illustration below, Anterior spines) and a few other physical characteristics. Two of these species, the marsh grass shrimp and the daggerblade, can also be recognized by the color of their eyestalks; the marsh grass shrimp's eyestalks are red-brown in color while the daggerblade's are yellow. These two species share a similar reproductive season, though in some areas, the marsh grass shrimp begins breeding two to three weeks after the daggerblade.

During spawning, the male marsh grass shrimp (*P. vulgaris*) approaches a recently molted (shed) female and touches her with his antennae. If she is receptive, the pair mate and she later releases her eggs in a steady stream from both of her oviducts. Held in place by stalks connected to her swimmerets, the eggs are continually ventilated by the movements of these appendages. The tiny larvae's compound eyes gradually become clearly visible through the eggs. Then, after some 15 to 20 days, the creatures begin the struggle to position themselves for hatching. The inner membrane of the two-layer eggs swells rupturing the outer membrane, and the larvae puncture the inner membrane using a sharp edge on the back of their abdomen.

Aided by the mother's ventilating movements, the larval shrimps are washed into a realm where they must quickly find food

Daggerblade shrimp, *P. pugio*

Marsh shrimp, *P. vulgaris*

Brackish water shrimp, *P. intermedius*

After Williams, 1984.

Anterior spines: The species of grass shrimp can be determined by the look of the anterior spine (rostrum).

or die. Within one to two days of hatching her brood, the female extrudes more eggs. Breeding is generally continuous throughout the summer, and in the southern part of their range, females hatched from a spring spawn often reproduce later that same summer.

The newly hatched larvae pass through some 10 stages (each stage = one molt) before completing larval development. As they drift, they feed on nearly anything that they come across. Their survival appears to depend on what they eat. In the laboratory, larvae that were fed only unicellular algae died within a short time. Those that were fed an all animal diet molted and grew more rapidly than those kept on a mixture of plant and animal.

Larval daggerblade shrimp
Redrawn from Broad, 1957.

The adults are also opportunistic foragers. They consume a variety of plants, animals and detrital matter. Research on the feeding habits of the daggerblade grass shrimp (*P. pugio*) has revealed the crustacean's special role in salt marsh environment. As it grazes on dead cordgrass leaves and stems (*Spartina* detritus), the shrimp breaks the pieces up into even smaller fragments, leaving them with rough edges. Within a short time, the roughen surfaces are colonized by bacteria and small diatoms. The daggerblade shrimp also excretes large quantities of ammonia and phosphate and dissolved organic matter. These materials (in a dissolved state) can be taken up by the diatoms and bacteria on the detritus. The resulting mass (detritus and its bacteria and diatoms) is richer in protein than the original detritus. The enriched detritus along with the shrimp's organic-rich feces are then consumed by the shrimp themselves and other detritus feeders.

During blooms of sea lettuce (*Ulva* spp.) and hollow green weed (*Enteromorpha* spp.), a site can become chocked with the decaying algae, robbing local waters of life-sustaining oxygen. Grazing by the daggerblade apparently helps prevent such accumulations. Its ability to function in a low-oxygen environment, an area that is generally uninhabitable to most predators or other competitors, is believed to allow the shrimp's population to grow. It thus fills an important niche in the saltmarsh ecosystem. ■

Female daggerblade shrimp, *Palaemonetes pugio*, with eggs. Mystic, CT.

Sand fiddler crab, *Uca pugilator*

Habitat: Sand or sand/mud with sparse vegetation. Burrows located at or above the high tide line, bordering salt marshes and tidal creeks.

Other common names: Fiddler, calling fiddler.
Phylum: Arthropoda. Subphylum: Crustacea. Class: Malacostraca.
Order: Decapoda. Family: Ocypodidae.
Geographic range: Cape Cod, MA, to Corpus Christi, TX.
Identifying characteristics: Male, a small claw and a large claw. Palm
    of the male's large claw is smooth. Female, two small claws.
Burrows: Temporary burrow with steep angle of descent (79°).
    Breeding burrow with shallow angle of descent (41°) and
    an expanded inner chamber.
Reproductive season: Long Island, NY - July to mid-August.
               Virginia - March to mid-summer.
               Miami, FL - April to October.
Waving behavior: Normal rate 6-10/min., stimulated 40/min.
Egg color: Black at time of attachment; grey a few days before hatching.
Egg production: .8-inch (20 mm) width female produces about 15,000 eggs.
Larval period: Six to eight weeks.
Life span of temperate water fiddlers: Seldom longer than 2 years.

*I*t is part of the creature's mating ritual. The male sand fiddler crab stands near the entrance of his burrow at low tide and waves his claw slowly back and forth hoping to attract a mate. If a female approaches, he increases the tempo of his waves to a frantic rate, and as she comes even closer, he begins to rap the ground with his large claw. When conditions are right, the female follows the male into his specially dug nuptial burrow. He then plugs the entrance with sand and the pair retreat to the inner chamber; unlike many other species of crab, mating takes place with the female's shell in a hardened condition (see Green crab, page 103).

After mating, the male sometimes digs another chamber, unplugs the burrow and begins courting again. While defending his residence he may mate with up to three females and provide them each with their own separate inner chamber. But a female is not always receptive to the male's

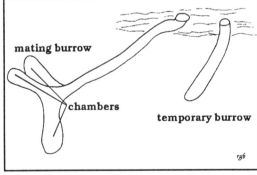

Sand fiddler crab burrows. After Christy, 1980.

19

Male mud fiddler, *Uca pugnax*, crab waving for a mate. Shea Island, CT. Long Island Sound.

seductive call. At times she enters the burrow only to make a hasty exit. She then stands just a few inches from the entrance. Her actions may initiate what is called the "Dash-Out-Back display" by the male.

As part of this behavior, the male bolts from the burrow and dashes off some 30 inches away. He raises his large claw as high as he can and, like a knight of old, he lowers it and charges back toward the burrow. Stopping just short of the entrance, he waves or raps it on the ground. If that does not entice the female, he sometimes simply shoves her down into his lair. At other times, the excited male loses sight of his burrow and runs right past it. Lost, he attempts to enter another male's residence with predictable results. The battle that ensues generally ends with the burrow's resident the winner. As will later be seen, however, if the intruder is successful in dislodging the occupant, it is usually because he is the larger of the two.

The male's courting activities are not confined to a daytime low tide; he also tries to attract a mate after dark. During a nocturnal low tide, however, he limits the activity to rapping on the ground. Both sexes detect the acoustic signals with vibration receptors (Barth's organ) located on each of their walking legs. When a male begins to wave or rap in response to an approaching female, the neighboring males usually join in a chorus of competition by increasing their own rate of waving or rapping.

The combative sand fiddler well deserves of its name, *U. pugilator* (*pugilator*, L. fist fighter). When a male encounters a neighboring male (or sometimes a female), he raises his large claw

20

and extends it somewhat to the side in a threatening posture. The claw is usually held in that position until the other moves away. If that doesn't drive off the creature, the male extends his claw further and flexes it in an arc; the encounter often ends at or before that point. But if a wandering intruder approaches with aspira-

Female sand fiddler crab, *Uca pugilator*, Shea Island, CT. Long Island Sound.

tions of taking over the burrow, the threats can quickly escalate into a battle. One of the combatants leads the charge and claws are locked together. In the melee, one of the two is occasionally lifted off the ground and flipped to the side. The winner takes over the burrow. If a wandering intruder is large enough and is sufficiently determined, he reaches into the burrow and pulls out its owner. If that doesn't work, he sometimes attempts to dig out the occupant, spending up to 40 minutes in the effort.

A wandering intruder uses at least two other types of behavior to take over an occupied burrow. In one strategy, he simply takes over a residence while the owner is some distance away. At other times, however, the intruder is much more deliberate. As he approaches an occupied burrow, he lowers himself to the ground by stretching out his legs. Coming even closer, he generally turns sideways so that only his small claw is visible; the fiddler recognizes the opposite sex by the presence or absence of a large claw. If the burrow's occupant does not challenge the intruder, the latter makes himself as inconspicuous as possible and lies in wait some 5 to 6 inches from the entrance. The instant the owner moves away, the intruder rushes in and a battle for ownership of the burrow invariably begins.

Other species of fiddler crab are also confrontational, but *Uca terpsichores*, a resident of the tropics, tries to clear the neighborhood of all male competitors. The aggressor repeatedly attacks a nearby competitor in an attempt to dislodge him. If he succeeds, he fills in the entrance to the abandoned burrow. When the competitor cannot be driven off, the aggressor resorts to building a large concave, arching structure (hood) or a barrier between them. He can then more easily defend his territory and have additional time to spend attracting a mate.

21

Despite all of the male competition, however, females of various fiddler species can also be quite selective in choosing a mate. The female sand fiddler's choice does not seem to be based as much on the male's size than on the quality and location of his burrow. The structure's integrity apparently helps assure reproductive success. During mating, attachment of eggs and incubation, the female remains in the burrow's inner chamber. When the structure is located too low in the intertidal zone, flooding may cause it to collapse. If, at that time, a female is in the process of attaching (oviposit) her eggs, she will fail and lose her brood. When an incubating female is forced to leave the burrow before the eggs are ready to hatch, she must find or dig a new burrow. Under those circumstances, she becomes more susceptible to predation.

After incubating her eggs for approximately two weeks, the female emerges on a nocturnal neap tide and makes her way to the water's edge. During the next two hours, she forcefully contracts her abdomen and dispatches her larvae in the receding waters. The timing of the release may help her, as well as her offspring, avoid predation.

The first larval stages resemble a mosquito (zoea). As they drift in the plankton, these tiny predators capture other animal plankton (zooplankton) by lashing their tail forward and pinning the unfortunates down with their tail (abdomen). The victims are then moved forward to the mouthparts. Over some six to eight weeks, the larvae grow and shed their body shell (exoskeleton), passing five zoeal stages and a crab-like megalopa stage. At that point, the creatures metamorphose into tiny bottom-dwelling crabs that are too weak to burrow but are able to cling to the substrate. After further development, they begin to burrow in the wet intertidal area. They later move to the area that is occupied by the adults, near or above the high tide line.

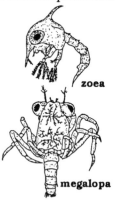

zoea

megalopa

Sand fiddler larvae.
From Bachand, 1979

The sand fiddler burrows in sand/mud mixtures though it tends to prefer sandier soils. Using the walking legs opposite its large claw, the male excavates his residence and deposits the resulting materials in small piles near the entrance or at some distance away. When the tidal waters begin to rise in the burrow, the crab plugs the entrance by pulling in sand from above and packing it up from below. Both sexes remain in their residence until the waters recede. They then emerge to feed or find a mate. Their activities are thus said to follow tidal and day/night rhythms; the

22

color of their shell is also affected by these rhythms.

As the light fades in the evening sky, the crab's shell changes from dark to a paler color. Within its external skeleton, specialized star-shaped cells (chromatophores) house pigment granules. When the granules are concentrated in the center of these cells, the creature's shell is light; when they are dispersed, its color is dark. The extent of the color change, however, appears to be also affected by exactly when the burrow is exposed in the receding tide. A sand fiddler whose burrow is uncovered for the longest, tends to show the greatest color change.

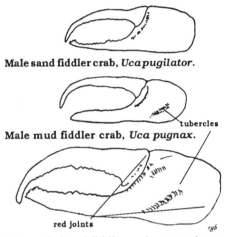

smooth palm (no turbercles)

Male sand fiddler crab, *Uca pugilator*.

Male mud fiddler crab, *Uca pugnax*.

tubercles

red joints

Male redjointed fiddler crab, *Uca minax*..
After Grimes, 1988

A foraging fiddler responds quickly to the rhythm of a falling tide. Joining many of its kind, it moves to the water's edge and begins to pick through the mud/sand sediments. The creature needs sand to feed itself; when experimentally restricted to the mud of a low marsh, the sand fiddler was unable to survive. The male scoops up sand grains with its small claw (both claws in the female) and using its specially adapted mouthparts, it scrubs food matter from them. Then light organic materials are ingested, and the heavier inorganics are periodically discarded as small balls. (These disposed of materials are easily distinguished from fecal pellets, which are rod-shaped.) The crab continues to feed at the water's edge until about mid-flood tide. Then, using celestial clues and landmarks, the crab finds its way back to its habitat.

The sand fiddler shares its temperate coastal environment with two other species of fiddler, the Atlantic marsh (mud) fiddler crab, *Uca pugnax*, and the redjointed (brackish water) fiddler crab, *Uca minax*. The marsh fiddler, the most common of the three species, inhabits mud or sand/mud substrates. Its burrow is sometimes found mingled with that of the sand fiddler. The redjointed fiddler, the largest of these crabs, is more tolerant of low salinity than the other two. It burrows in mud, mud/sand or a dense root mat that is frequently flooded by fresh water.

Say mud crab,

Dyspanopeus (=Neopanope) sayi

Habitat: Mud and oyster shell bottoms.
Low intertidal, under rocks.

Other common names: Mud crab, black-fingered mud crab.
Phylum: Arthropoda. Subphylum: Crustacea. Class: Malacostraca.
Order: Decapoda. Family: Xanthidae.
Geographic range: Nova Scotia to Florida Keys.
                     Most common mud crab north of Delaware Bay.
Depth range: Low intertidal under rocks to 150 feet (46 m).
Reproductive season: Virginia - May to September.
Egg production: Approximately 100,000 over its lifetime.
Larvae found in plankton: In LIS, June to October.
Life span: 2 to 3 years.

*T*his pugnacious little crustacean is usually less than an inch wide, yet, armed with an impressive-looking claw, it can easily crush barnacles, young oysters and half-inch quahogs. When threatened by an inquisitive human, it pinches without mercy and can cling tenaciously to the fleshy part of a hand for several excruciating minutes. Struggling with the nasty little critter frequently makes it squeeze even harder! Closely resembling other regionally occur-

Say mud crab, *Dyspanopeus sayi*. Bridgeport, CT. Long Island Sound.

24

ring mud crabs, the Say mud crab, (named for Thomas Say) *Dyspanopeus sayi*, is the most common of its kind residing in Long Island Sound. All are related to the delicious Florida stone crab, *Menippe mercenaria.*

Dyspanopeus can be very aggressive with its own kind, but it usually avoids confrontation. When approached by another mud crab, the two ordinarily detour around each other. If they come into contact, one or both retreat. The encounter becomes more heated when an individual's claw touches the other's, or if both of a crab's claws are stretched out horizontally in what is known as a Lateral Merus display. At that point the pair charge each other, propelling themselves in a rapid sideways gait as they grab and pinch. The fights generally only last a few seconds, and the crabs retire without visible damage. Preparation for mating often resembles such a confrontation.

Unlike many other species of crab, the Say mud crab mates while the female's shell is hard (see Green crab, page 103). Either sex may initiate the encounter but as the ritual progresses, the male becomes more aggressive. At some point, he lunges toward the retreating female with his two claws outstretched, and takes her within his grasp. For the next few seconds, the pair remains motionless and if the female is receptive, she keeps her claws tucked in close to her body. The male then begins to twitch the ends of his walking legs, and the female follows suit. Using his claws, the male positions the female underneath himself and brings their undersides in contact; their abdomens are hinged backwards and mating takes place.

Mating couples form no permanent bond. During the season, both sexes probably mate several more times with different partners. In her first reproductive year, a 0.31-inch (8 mm) wide female sheds four to five times. At that size, she produces about 2,400 eggs. Each time she sheds, she grows in width, mates and produces more eggs. By the end of the season, the female has produced 15,000-30,000 eggs with a total of about 100,000 over her lifetime. Very few of the crabs live to the middle of their

Say mud crab.

Flatback mud crab,
E. depressus

After Williams, 1984.

Major claws: Major claws of two common mud crabs. The Say mud crab is the most common of its genus in Long Island Sound. The flatback mud crab is very common in Chesapeake Bay.

Laval Atlantic mud crab,
Panopeus herbstii.

Panopeus is a close relative of other regional mud crabs including the Say mud crab.

25

third summer. During their short lives, however, they are important shellfish predators and they serve as food themselves for a variety of fish and seabirds.

When a mud crab has located a buried young northern quahog (*Mercenaria mercenaria*), it digs it way down to its prey. It then cracks the clam open and often consumes the mollusk while both are still buried. When offered small quahogs in a laboratory setting, the crab consumed up to 14 clams in an eight-hour period.

In Great South Bay, NY, the mud crab is the most important predator of juvenile quahogs; apparently it can survive on a diet of the clams alone. It is, however, not the only predator of the mollusk. The list of the clam's other crab-predators includes the rock crab, blue crab, lady crab and green crab. Sea stars and predatory snails (eg. moon snails, oyster drills, whelks) also feed on the clam. If an area is inhabited by a large number of predators, nearly all of the quahogs are consumed. Only those that settle near rocks or other solid objects can be expected to survive.■

Blue crab, *Callinectes sapidus*

**Habitat: Tidal creeks, rivers and sounds.
Variety of bottoms.**

Other common names: Blue claw, soft-shelled crab, peeler, Sally crab (young female), sooks (adult female), jimmy, channeler (adult males).
Phylum: Arthropoda. Subphylum: Crustacea. Class: Malacostraca.
Order: Decapoda. Family: Portunidae.
Geographic range: Nova Scotia to northern Argentina. North Sea,
                  France and northern Adriatic.
Depth range: Usually inhabit the shallows to 120 feet (37 m).
Salinity tolerance: Freshwater to hypersaline lagoons (44-48 ppt).
Reproductive season: Chesapeake Bay - May through October.
Egg production: 700,000 to 2 million eggs, each about .25 mm in diameter.
Incubation period: One to two weeks.
Larvae: Approximately seven to eight zoeal stages and a megalopa stage.
Size increase with each molt: 25 to 40%
Maximum size: Measured from tip-to-tip of the spines, males usually to
              8.22 inches (209 mm); Females to 8 inches (204 mm).
Life span: Approximately 3 years.

*T*his beautiful and graceful swimmer produces some 700,000 to 2 million eggs. Yet, despite the astronomical numbers, only one in a million of the blue crab's brood survives to become a mature adult! Just what causes these tremendous losses can be at least partially understood by examining what is known of the creature's life history.

26

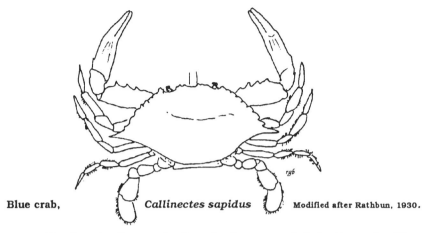

Blue crab, *Callinectes sapidus* Modified after Rathbun, 1930.

Similar to the majority of other coastal, bottom-dwelling invertebrates, the blue crab's larvae are released into the water where they drift with other planktonic creatures. The larvae shed (molt) their external skeletons as they pass through approximately seven to eight mosquito-resembling stages, and a crab-like megalopa stage (see Green crab, page 103). They then metamorphose into tiny bottom-dwelling crabs, and following 18 to 20 more sheds, the juveniles reach sexual maturity. At that point, they are approximately 1 to 1.5 years of age. The male continues to shed his shell and grow after maturity, but the mature female generally ceases shedding and growing. Her final shed is known as her terminal or pubertal molt.

In preparation for mating, the female develops sperm storage sacs (spermathecae) and her abdomen changes dramatically (see illustration, Blue crab, abdomen). As the time for her terminal molt draws near, she releases a chemical substance (pheromone) in the water that helps attract a potential mate. If interested, the male may approach her with his claws outstretched, and his legs extended so that he walks high off the sea floor. He also raises and waves his paddle-like swimming appendages from behind his shell.

When the male shows no response, the female frequently bumps up against him. Ultimately, her advances or his behavior, with or without a courtship display, leads to the male's role as a guardian. Picking the female up with his first pair of walking legs, the male carries her with her back held against his abdomen in what is known as a cradle-carry or hard-doubler. She is held in that position for about two days and is released during shedding. Mating takes place within a short time of the shed. Still "soft," the female is picked up and held with her abdomen facing his; fishermen refer to this

27

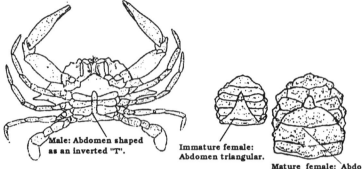

Male: Abdomen shaped as an inverted "T".

Immature female: Abdomen triangular.

Mature female: Abdomen broad and rounded.

Blue crab, abdomen: underside (ventral).
From Hill, 1989.

stage as a soft-doubler. Over the next five to 12 hours, the male directs his sperm into the female's storage sacs with his stylus-like appendages. The sperm remains viable for up to a year and is often used for two or more spawnings. Following mating, the male continues to protect the female by carrying her in an upright position until her shell is sufficiently hard.

Reproductive activities generally take place in brackish water (low salinity) at the upper reaches of an estuary. The female then migrates to higher salinity waters, at the mouth of the estuary or deeper offshore. The male usually remains in the low salinity waters for the rest of his life, moving to deeper water during the cold of winter. Some two to nine months after mating, usually in the spring or summer, the female extrudes her eggs and carries them as a large mass (sometimes called sponge) under her shell. As they incubate over a period of one to two weeks, absorption of the yolk and the larvae's developing eyes change the color of the eggs from orange to yellow, to brown and then to black.

Predation of the brood begins immediately. The eggs are frequently attacked by small fishes and a nemertean worm, *Carcinonemertes carcinophila*, which inhabits the crab's gills and invades the egg mass. The unhatched eggs may also succumb to fungal infections, pollutants, stagnant water and unusually high temperatures. After being released into the offshore plankton, the crab's larvae are preyed upon by fishes, jellyfishes, combjellies and a variety of other plankton-eating creatures. As they drift and mature, the larvae must find their way back

Blue crab larva, zoea.
Redrawn from Costlow, 1959.

28

to shore or they perish. At the mouth of the Delaware Bay, researchers have found the crab-like megalopa larvae near the surface in much greater numbers during a flood tide, than on a falling or ebb tide. These findings suggest that the larval crabs ride the currents into the bay during a flood tide. To avoid being swept back out, they move toward the bottom during ebb tide. In Chesapeake Bay, the megalopae return to the bay with the aid of northerly winds that occur during the late

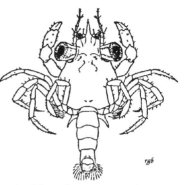

Final larval stage, megalops.

summer and fall. While studying the bay's population of megalopae in 1985, scientists noted that the Hurricane Juan's storm surge carried in large numbers of the larvae. Such occurrences or lesser events that help move the larvae shoreward, may explain sudden increases in crabs during the following year. A poor harvest of blue crabs is likely to occur when storms or unfavorable currents prevent the larvae from returning to shore.

Shortly after the megalopa larvae enter the estuary, the creatures metamorphose into tiny crabs. They then continue their migration toward the less-saline areas. Cold weather and heavy rains that dramatically lower salinity may affect survival of the young crabs. Predation, however, continues to be a major factor in their survival (as well as human impacts on the creatures - pollution, loss of habitat and overfishing). The long list of predators includes striped bass, spotted sea trout, black drum and the American eel. Certain seabirds also feed on the juvenile crabs. In the turbid waters of coastal Texas, small crabs (less than 25/64 in., 10 mm) have apparently learned to flee or hide at the sound of the laughing gull, *Larus atricilla*. Despite the gauntlet of difficulties faced by the larvae and juveniles, a few nonetheless manage to reach sexual maturity and become active predators in their environment. In this role they can impact commercially important species.

The blue crab is considered a scavenger, but it also feeds on a variety of plant materials, algae and detritus. Its prey includes fishes, crabs, mussels, clams, oysters, snails, tunicates and even jellyfishes. When foraging for the soft-shelled clam, *Mya arenaria*, it is assumed that the crustacean finds the mollusk by using chemical and tactile cues. Like other crabs and lobsters,

What's in a word?
Callinectes sapidus
Calli (Gr. kalos) = beautiful
nectes (Gr. nektes) = swimmer
sapidus (L.) = tasty

29

the creature has specialized odor-detecting (chemoreceptive) organs located on its first antennae (shortest antennae, antennules). In the blue crab, the odor-detecting organs appear as a small tuft of hair-like structures on the outer branch of the antennae. The hair-like setae on the crustacean's walking legs are also used for chemoreception.

When the crab has located a clam, it prods the bottom with is walking legs and then begins to dig it out. The creature excavates by pushing sediments with its claws and legs, and occasionally carrying away materials in the crook of its claw. Once it is able to grasp the clam, the crab holds it with its claws and cracks its victim open with its other claw. Even small crabs are able to get to the clam's flesh by first chipping away at the shell's weaker edges.

Bottom-dwelling fishes can also be easy prey for the blue crab. Lying motionless and sometimes partially buried, the crustacean waits patiently for a prey; the naked goby (see Naked goby, page 122) seems oblivious to the danger. If the tiny fish strays to within reach, the predator snaps it up, often pinning the goby up against its shell. Unable to escape, the fish is quickly devoured.

Man the predator also plays a role in this realm of the hunter and hunted; the crab is taken for human consumption and for bait. Throughout its range, the blue crab is sought after by recreational fishermen. In areas of abundance, it is also harvested by commercial fishermen. The crustacean is the object of the country's largest crab fishery, accounting for about half of the total weight of crab landings. Along the Mid-Atlantic States, most of the commercially caught blue crabs are taken from the Chesapeake Bay region, in Maryland and Virginia. A smaller number are taken from New Jersey and Delaware. During 1984, landings of the crustacean in the Gulf of Mexico exceeded 51 million pounds. Though most of the animals are taken to market with their shell hard, a small percentage are sold as recently shed, soft-shell crabs. Because of the difficulty of maintaining them in this state, they bring a higher price.

Soft-shelled crabs are captured shortly after having shed or as peelers -crabs that are preparing to shed. As the new shell develops and darkens, it becomes visible, especially through the partially transparent last two segments of the swimming legs. Baymen use the color changes to estimate time of shedding. A white-rim crab (green-sign crab) generally sheds within one to two weeks. A pink-rim or pink-sign crab sheds in three to six days, and a red-rim or red-sign crab sheds in one to three days. A "buster" is a crab that has already begun to rupture and shed its old shell.

Mantis shrimp, *Squilla empusa*

Habitat: Mud, mud/sand, silt/clay.

Other common names: Common mantis shrimp, snapper shrimp.
Phylum: Arthropoda. Subphylum: Crustacea. Class: Malacostraca.
Order: Stomatopoda. Family: Squillidae.
Geographical range: Cape Cod, MA, to the Gulf of Mexico.
                    South to Surinam.
Depth range: From the shoreline to 505 feet (154 m).
             Generally in less than 131 feet (40 m).
Diameter burrows: .4 to 4 inches(10 to 100 mm).
Normal size range of adult: 6 to 8 inches (15 to 20 cm).
Size of sexually mature female: 2.8-3.1 inches (70 to 80 mm).
Reproductive season: Long Island Sound, July to October.
                     Chesapeake Bay, July to October.
                     Gulf of Mexico, January to July-August.
Size of eggs: Approximate diameter, 1/3 mm.
Egg production: Large female, in excess of 100,000 eggs.
Life span: Uncertain. Estimates of 5 to 6 years.

*T*he mantis shrimp lies-in-wait at the entrance of its burrow with only its periscope-like eyes and antennae extending above the rim. Unmindful of the danger, a small fish swims into range and the hunter strikes. Thrusting its spine-armored limb forward, the mantis spears its victim and withdraws the appendage. The strike is complete in just four to eight milliseconds; "it is the fastest animal movement known" (Caldwell, 1976).

The predator devours the creature with equal gusto. Using the grinding surfaces on its paired jaws (mandibles) and the hooks on the ends of its third, fourth and fifth limbs or thoracic appendages (maxillipeds), the mantis shreds and consumes its prey. Roy Caldwell, the dean of the natural history of these unusual crustaceans, recounts how a 10-inch long Indian Ocean mantis dispatched a 4-inch fish in just four minutes!

Worldwide, there are some 350 known species of mantis shrimps or stomatopods. Ranging in size from 1/2 to 15 inches (15 to 380 mm), most live in tropical and subtropical waters. Stomatopods are generally divided into two groups according to the form and function of their spearing or smashing (raptorial) appendages. The spearers, as the name implies, impale prey. The smashers are equipped with raptorial appendages whose heels are greatly enlarged; they break apart their victim's shell or carapace. One species of smasher, which is 10 inches (250 mm) long, is said to produce an

31

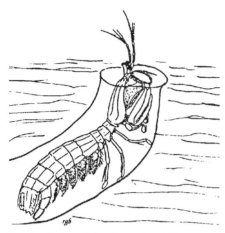

Mantis shrimp lying-in-wait at the entrance of its burrow.

impact nearly equal to a small-caliber bullet.

The spearers tend to live in burrows they dig for themselves. Most smashers, however, compete for existing burrows or hollowed-out areas found in rocks, corals and calcareous algae. Along the Atlantic coast and into the Gulf of Mexico, the spearing mantis, *Squilla empusa*, digs its burrow in stable mud, silt/clay or compact sand. It is one of only a few successful temperate water stomatopods.

At the northern end of its range, *Squilla empusa* digs two types of burrows; one is used during the summer and the other serves as a winter refuge. The burrow is created using a combination of carrying excavated material, bulldozing and fanning movements of its paddle-like swimmerets (pleopods) to blow away loose sediment. As work progresses, the mantis digs head-first into the bottom and then pushes or caries the tailings to the surface and spreads them out around the shelter's opening. During the summer, the creature produces a U-shaped tunnel with an opening at both ends. Dug to a maximum depth of about 1.5 feet (0.5 m), its diameter is sufficient to allow the animal to reverse direction within its confines. SCUBA divers can capture this local mantis by reaching in through one entrance with a "gloved hand," and blocking the exit with a large fish-tank net. That's the easy part. Picking up the feisty creature with its sharp raptorial appendages and its spine-armored tail (telson), is quite another. It must obviously be handled with care. But a chance to closely examine this fascinating crustacean is well worth the effort.

In Georgia, *Squilla empusa* remains year-round in a summer-type of burrow, but in Rhode Island and Long Island Sound, it must somehow protect itself from the winter's cold. Unable to withstand temperatures of approximately 41°F (5°C) or less, the mantis excavates a 13-foot (4.1 m) deep burrow that is nearly straight down. It then spends the season at the bottom of the shaft where it shows little or no perceptible activity.

As the waters warm, the creature leaves its winter refuge and prepares itself for its reproductive season. Fertilization is internal and sometime after copulation, the female extrudes her eggs.

Forming them into large circular sheets, they are eventually wadded into a ball. The female moves her brood around in the burrow with her front limbs, aerates them and keeps them clean using her mouthparts. In Long Island Sound and Chesapeake Bay, *Squilla's* larvae first make their appearance in the plankton during July; in the northern Gulf of Mexico, they first appear in January.

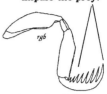

teeth or spines that impale the prey.

Mantis shrimp's spearing claw.

Unlike the larvae of many other species of invertebrates, *Squilla empusa's* planktonic offspring somewhat resemble their parents; they have the large raptorial appendages of the adults. The larvae are voracious predators that feed on other zooplankton as large as themselves. As they drift in the currents, the creatures shed their outer skeleton. They pass through nine stages and attain a length of about 11/16 inch (17.5 mm) before metamorphosing to a bottom-dwelling existence. The predatory larvae are also prey. In tropical waters, stomatopod larvae represent a significant portion of the zooplankton population. They are an important part of the diet for certain reef fishes, jacks, snappers, scads, herring, mackerel and tunas.

The adult *Squilla* is a bottom-dweller, but it is also a powerful swimmer. Using a fanning motion of its five pairs of swimmerets, the mantis may sometimes be seen propelling itself in a jerky, looping pattern. In Narragansett Bay, RI, and at Hope Bay, MA, the creature has been observed swarming in large numbers near the surface during late October to December. The behavior may be related to reproduction, migration or preparation for winter burrowing.

When outside its burrow, the mantis often stands on the outermost edges of its tailfan (telson), the tips of its swimmerets and the last three of its eight pairs of limbs. Its first pair of limbs, or maxillipeds, are used in grooming. Maneuvering them much as a small pair of gloved hands, the mantis carefully cleans its eyestalks and antennae with bristles that are specially adapted for grooming those appendages. Other bristles are used for the carapace and general cleaning. To reach the rear parts of its body, the animal rolls itself nearly into a ball. Though *Squilla empusa* can be kept in an aquarium, it often dies in captivity during the shedding (ecdysis) of its shell (See Green crab, page 103). The animal is considered edible, but due to its size, its has not been commercially exploited like its close relative, *Squilla mantis*, in the Adriatic Sea. ■

The fishes

Mummichog, *Fundulus heteroclitus*

Habitat: Shallow brackish coves, inlets and tidal marshes.

Other common names: Chub, mud minnow, pike minnow,
marsh minnow, gurgeon, mummy, killifish.
Phylum: Chordata. Class: Osteichthyes.
Order: Atheriniformes: Family: Cyprinodontitae.
Geographic range: Gulf of St. Lawrence to northeastern Florida.
Gulf of Mexico - Texas (closely related species).
Depth range: Shoreline to about 12 feet (3.7 m).
Size: 2-4 inches (51-102 mm), maximum 6 inches (152 mm).
Reproductive season: Chesapeake Bay, April to September.
Long Island Sound, May to July.
Life span: 2 to 3 years, occasionally to 4.

*T*he rising tide bathes the egg for no more than a moment when activity begins. The egg swells, the embryo stretches its mouth as if awakening from a long sleep, and its heart rate and respiration increases. A special chemical (an enzyme - chorionase), released from glands in the creature's mouth and gills, weakens the egg's wall. Within 15-20 minutes a tiny larval fish breaks free and escapes in the receding tide.

The mummichog's spawning begins under the light of a full or new moon. As the rising waters of a spring tide near their peak, females laden with eggs signal anxious suitors by turning on their sides and displaying their silvery-white undersides. Competition for mates is intense and when individuals finally pair up, the males hold themselves with the aid of their fins, in intimate contact with the females. Propped up against objects such as the stalk of salt marsh cordgrass (*Spartina alterniflora*), the pairs quiver and expel streams of eggs and sperm onto the inner surface of cordgrass leaves or on empty ribbed mussel shells. There, the 100-plus eggs from each spawning couple remains to incubate in the open air until the next spring tide washes over them and triggers their hatching.

During the season, the nuptial ritual is repeated on as many as eight separate occasions; each time, it coincides with the spring tide of a full or new moon. Though the fish generally spawns at night, it may also spawn during the day.

The mummichog also feeds during daylight. Equipped with a protruding lower jaw and a mouth that are turned upward, the creature is perfectly adapted for feeding on the surface where it

34

consumes zooplankton, small fish and grass shrimp. As a predator, it is inclined to devour almost anything within its reach. On the sea floor, it feasts on polychaete worms, small bottom-dwelling crustaceans, snails, soft-shelled clams, algae and fish eggs (even of those of its own species).

Mummichog, *Fundulus heteroclitus.*

This voracious predator, however, is also prey for others, and so plays an important part in the salt marsh food chain. It is a source of food for herons, egrets, gulls, terns and other coastal seabirds. Juvenile bluefish, striped bass, summer flounder, American eel and various species of crabs also actively feed on the fish.

Determining the sex: Gender is easily determined by size and color. A 1-year-old male is smaller than the female, and its body color is dark green or olive with a yellow underside. From its midsection to its tail, the fish has some 15 pale vertical stripes, and in the same area, it is sprinkled with numerous yellow or white dots. During spawning season, its colors become more intense. The female's body tends to be a drab olive-green with a lighter belly.

The salt marsh is a difficult environment. Tides, heavy rains, and seasonal and daily temperature changes, combine to subject it to wide and sometimes precipitous changes in salinity and temperature. The ability of the mummichog to tolerate such changes seems to make it ideally suited for life in this ecosystem. In the laboratory, the fish adapts so quickly to dramatic changes in salinity that no ill effects have been demonstrated. Equally adept to fluctuations in temperature, it has been shown to recover from exposure to 104°F (40°C). In the cold of winter, the mummichog copes with the temperature extremes by becoming very sluggish and sometimes burying itself in 6 to 8 inches of mud.

This amazing little fish can also withstand low dissolved oxygen and high concentrations of certain pollutants such as No. 2 fuel oil. When oxygen levels drop sufficiently, the mummichog gapes at the surface. Marine scientists presume that the activity creates enough turbulence to oxygenate the water reaching the gills.

The chub, as it is also called, is widely used as a baitfish. It is, however, of equal importance to investigators who have used it for research in embryology, genetics, physiology, behavior and many other disciplines. During the mid-1970s, mummichog eggs were flown into orbit aboard the Apollo command module to observe the effects of prehatching weightlessness. ∎

Striped killifish, *Fundulus majalis*

Habitats: Tidal marshes, brackish ponds, creeks and rivers.

Other common names: Killie, banded killifish, striped mummichog,
                            mummie, bull minnow, mayfish, rockfish, gudgeon.
Phylum: Chordata. Class: Osteichthyes.
Order: Atheriniformes: Family: Cyprinodontitae.
Geographic range: Massachusetts to Florida.
Size: Record length of 8 inches (203 mm).
Reproductive season: New Jersey - northward June to August
                            Chesapeake Bay - April to September
Eggs: Spherical and transparent - 2 mm in diameter

*I*t is never far from shore. The striped killifish often travels in schools as it patrols the water's edge, moving in and out with the tides. In pursuing its routine, however, the fish is sometimes washed ashore or becomes stranded in the receding tide. When trapped in a shallow pool, the killifish has been observed swimming rapidly around the edges. It then launches itself to solid ground. In a series of well-directed jumps that can span a few inches to several feet, the fish propels itself back to open water. During one experiment, over 200 killifishes were observed flipping their way around the ends of a dam that was blocking the shallow outlet of a tide pool. As if following the leader, the fishes made their way, one after the other, across about 10 feet of dry ground before reentering the water. The mummichog, a close relative of the killifish, has also shown the ability to find its way back to the water when stranded.

There are many similarities between the two species. They often share the same habitat, and they tend to prefer feeding on small crustaceans and polychaete worms though they generally don't pass up any small prey. They are very tolerant of changes in temperature and salinity, but the killifish is less likely to enter freshwater than its counterpart; it also tends to be found in higher salinities. Both species spawn in the shallows.

The spawning season for the killifish begins as early as April in Chesapeake Bay and in June from New Jersey northward. Large schools of fishes assemble along the shoreline and individuals pair off. In a ritual that closely resembles the mummichog's reproductive behavior, the male clasps the female with their bodies held is an S-shaped curve. Reproductive products are then released. Unlike the

36

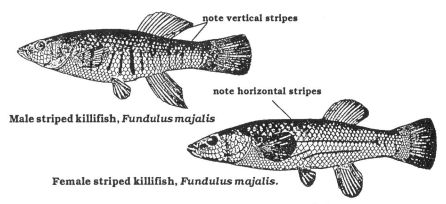

note vertical stripes

note horizontal stripes

Male striped killifish, *Fundulus majalis*

Female striped killifish, *Fundulus majalis*.

From Abraham, 1985.

mummichog, however, the killifish often buries her eggs near the shoreline, during a low tide. The killifish's eggs are sometimes found deposited alongside those of the horseshoe crab.

It is generally difficult to mistake an adult male killifish for a female. The mature female has one or more dark longitudinal lines on its sides and one or more vertical stripes near its tail; a male carries 15-20 black vertical stripes along his sides. The young of both sexes also have vertical stripes, but these total no more than seven to 10; the sex-specific pattern does not appear until the fish has grown to between 1 and 2 inches.

There have been occasions, however, which despite the size and pattern, the sex of the fish was still uncertain. Outwardly appearing males have been found carrying eggs. These same individuals later produce milt. It should not be surprising, however, since in the normal course of their life cycle, other species such as the black sea bass and various parrotfish are known to begin their lives as functional females and later become males (see protogyny, Black sea bass, page 137). Some invertebrates such as the eastern oyster follow the opposite tact; they begin as males and later become females. The silverside, however, has its own unique sexual strategy (see Sliverside, page 40). ■

Similar to its close relative, the mummichog, the striped killifish can be easily captured in unbaited or baited minnow traps. Both species adapt well to an aquarium, tolerating even poor water quality. The fishes feed on almost anything.

37

Sheepshead minnow,
Cyprinodon variegatus

Habitat: Shallow bays, inlets and salt marshes.

Other common names: Variegated minnow.
Phylum: Chordata. Class: Osteichthyes.
Order: Atheriniformes. Family: Cyprinodontidae.
Geographic range: Cape Cod to Mexico.
Depth: Often found in less than 3 inches of water.
Size: To a maximum of 3 inches (76 mm).
Reproductive season: May to September.
Eggs: Spherical, 1-1.73 mm, adhesive, sink to the bottom.
Incubation period: Five to six days.

*T*he striped killifish moves its tail almost imperceptibly as it edges its way toward the sheepshead minnow. When it has approached to within a few inches, it assumes a head-down position and holds itself in that posture by rapidly beating its pectoral fins. The sheepshead then positions itself alongside its client and begins to clean it of external parasites.

The cleaning symbiosis is a behavior that is usually associated with residents of tropical seas. Yet, the sheepshead minnow and other temperate water fishes have often been observed acting as cleaner fish in aquariums. It is widely believed that they regularly do so in the wild.

Generally only little more than an inch long, the feisty sheepshead does not hesitate to chase or physically attack others many times its size. But the soliciting posture of the killifish alters the minnow's behavior and it readily takes to grooming its host. During this mutually beneficial process, the killifish twitches frequently

Sheepshead minnow

Striped killifish

Striped killifish presenting itself for cleaning in a head-down position.
After Able, 1976.

38

Unfertilized egg with attachment filaments

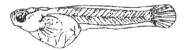

Newly hatched sheepshead minnow, 3.4 mm.

Determining gender:
The adult female sheepshead minnow has black stripes with a black dot on the end of her dorsal fin; the adult male has no stripes. During spawning, the sides of the male become an iridescent blue-green and his underside yellow. The young of both sexes have black bars on their sides.

and sometimes swims a short distance off, but it immediately returns to resume its head-down posture. It remains there until one or the other breaks up the partnership by swimming away. The sheepshead then resumes its aggressive behavior.

During the spawning season, the adult male sheepshead is especially ill-tempered. The fish establishes a nest in the shallows and guards the site from any intruder. When a female of its own species approaches its territory, she is driven off unless she is ready to spawn. When conditions are right, however, the male moves alongside his mate and holds himself firmly against her with his fins. Their bodies then shudder as eggs and sperm are expelled to the sea floor. In less than a minute, nuptial rights are repeated up to four more times. Though the female leaves the nest after spawning, she often lays eggs several more times during the season. The male remains at the nest and welcomes any other ripe female that approaches his territory.

The sheepshead minnow sometimes runs in large schools along the edges of tidal creeks, at depths of 1 to 2 inches of water. As the tide rises, it invades the flooded salt marsh cordgrass and feeds on vegetable matter and small invertebrates. It returns to open water when the tide begins to ebb. If the fish becomes stranded, it burrows into the soft mud or plant debris where it awaits the next flood tide. The minnow also burrows into the mud during the cold of winter.

The pugnacious little sheepshead is extremely tolerant of foul water. It is capable of withstanding water that is devoid of oxygen and it has been reported in sulfurous saline water. The creature is too small to be of commercial value, but its size makes it an ideal food for many species of predatory fish. It can be captured in a beach seine or minnow trap and is easily maintained in captivity. In the confines of an aquarium, however, the sheepshead often tends to bully others.

Other temperate water cleaner fishes: Mummichog, rainwater killifish and the four-spine stickleback.

Atlantic silverside, *Menidia menidia*

Habitat: Intertidal creeks, salt marshes, shoreline of bays and estuaries. Offshore during the winter.

Other common names: Capelin, shiner, spearing, sperling, green
                  smelt, sand smelt, white bait.
Phylum: Chordata. Class: Osteichthyes.
Order: Atheriniformes. Family: Atherinidae.
Geographic range: New Brunswick, Nova Scotia to Volusia County, FL.
Adult size: 4 to 5 1/2 inches (102-140 mm).
Reproductive season: Early May to June.
                         Peaks during mid-May in Long Island Sound (LIS).
Egg production: 4,500 to 5,000 in 4 to 5 separate spawnings.
Egg size: .09 to 1.2 mm in diameter.
Incubation time: 3 days at 30°C (86°F), 15 days at 18°C (64°F).
Life span: 1 year, occasionally 2 years.
                    Average of 13-15 months in LIS.

*T*he Atlantic silverside is a handsome, silvery little fish whose hidden iridescent blue and green colors can best be appreciated in the beam of a strong light. It is ordinarily one of the most abundant fishes along the shoreline of the coastal Atlantic marshes. But it will later be explained, the method by which nature determines the sex of its offspring is anything but usual.

Preparation for spawning begins during the daytime, generally about one hour before high tide. Large schools move within a few feet of the shoreline and maintain their position until just before the tide reaches its peak. At that moment, individuals begin to swim between flooded saltmarsh cordgrass. The females release their eggs near the base of the plants, among clumps of algae, or at times, in the empty burrows of sand fiddler crabs. They are then followed by one or more males who fertilize the eggs with their milt.

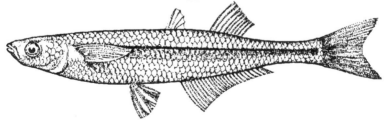

Atlantic silverside, *Menidia menidia*. After Bigelow, 1953.

40

Frantic spawning activity involving enormous numbers of the fish nearly deplete the area waters of oxygen. The phenomenon, however, works to their advantage. Predators that might attack the exhausted spawners are apparently unable to penetrate the oxygen-poor waters, and the silversides thus gain time to recuperate.

During the reproductive season, the females spawn three to four more times, each depositing a total

> **Determining gender:** The sex of an Atlantic silverside cannot be ascertained externally. Once the fish has grown to 3/4 inches (29 mm) in length, however, gender can be determined by examination of the gonads under a dissecting microscope.

of about 5,000 eggs. Their eggs, which are about 1 mm in diameter, are transparent with a yellow to green hue. Within them, there are five to 12 large oil globules and many more small ones. Though the eggs themselves are sticky and adhere to each other and to the substrate, their attachment is also aided by tiny filaments that extend from their outside surface. The eggs incubate for about three to 15 days, depending on water temperature. After hatching, water temperature continues to exert an influence by helping determine the sex of the larval fish.

Hatching generally occurs at night during a high tide, a strategy that probably helps the larval fish avoid predators. If the growing hatchlings are subjected to low temperatures, typical of early in the season, the majority become females. The warmer temperatures of early to mid-summer, produce more males. Over the course of the season, however, the ratio of male to female eventually evens out at one to one. At the northern reaches of the silverside's range where cold temperatures are constant, the ratio is also one to one. Thus, in that area, genetics and not the hatchling's environment, is the greatest determinant of gender. It is generally believed that the early development of more females than males allows the females to grow larger and thus increases their potential for producing more eggs.

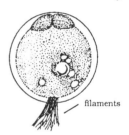

filaments

Silverside's fertilized egg. 1-1.5 mm in diameter.
The silverside's eggs sink to the bottom and adhere to each other, seaweeds, eelgrass, beach trash or sand.

From Hildebrand, 1928.

> **Incubation temperature and sex determination:**
> The sex of many reptiles is known to be controlled during incubation by temperature. These include many species of turtles, some lizards and all crocodilians.

Silversides are opportunistic feeders. They prey on a variety of tiny planktonic creatures that include crustaceans, newborn squid, worms, fish eggs and insects. They also feed on algae. From the

Atlantic silverside, *Menidia menidia*, Mystic, CT. (Photographed at night.)

time of hatching through mid-autumn in Long Island Sound, the young fish grow at a rate of about 25/64 to 39/64 inches (10-15 mm) per month. By late fall, they migrate into deeper water where growth all but ceases. Winter is difficult for the species, and as many as 99 percent perish. The survivors spawn at age 1 and usually die at the end of that season. Very few of the silversides survive through the following winter to see a second birthday.

The Atlantic silverside is an important part of the saltmarsh food web, providing forage for game fish and seabirds. It is used as a baitfish and during World War II, it was occasionally used for human consumption. ∎

> The fish can be captured in minnow traps or beach seines, but it is easily killed or injured during handling. Once established in a salt water aquarium it thrives on brine shrimp and occasional flake food.

The reptiles
Diamondback terrapin,
Malaclemys terrapin

From US Atlas, 1989

Habitat: Tidal marshes.

Other common names: Terrapin, Chesapeake terrapin.

Phylum: Chordata. Class: Reptilia. Order: Testudines. Family: Emydidae.

Geographic range: Cape Cod, MA, to Corpus Christi, TX (Seven subspecies).

Northern terrapin, *M. t. terrapin*: Cape Cod, MA, to Cape Hatteras, NC

Carolina terrapin, *M. t. centrata*: Cape Hatteras to northern Florida.

Florida east coast terrapin, *M. t. tequesta*: Atlantic east coast of Florida.

Mangrove terrapin, *M. t. rhizophorarum*: Florida Keys.

Ornate diamondback terrapin, *M. t. macrospilota*: Florida Bay to Alabama.

Mississippi diamondback terrapin, *M. t. pileata*: Florida panhandle to western Louisiana.

Texas diamondback terrapin, *M. t. littoralis*: Western Louisiana to western Texas.

Maximum size: Female to 10 inches (250mm) - rare; record weight 7 1/2 pounds. Male seldom longer than 5 inches (125 mm).

Reproductive season: Massachusetts - mid June to mid July.

New Jersey - early June to late July.

Florida - late April to early July.

Incubation period: Varies according to the temperature; 60-104 days.

Life span: May exceed 40 years.

*A*roused from her winter slumber, the terrapin prepares for the rites of spring. She swims to the quiet waters of a salt marsh with just her nose breaking the surface; propelled by her webbed-hind feet, a small wake reveals her presence. Then, under cover of night or the early morning hours, a male mounts the larger female and mating takes place at the water's surface or on the muddy bottom.

In time, the warmth of a mid-June suns beckons the egg-laden turtle to a narrow strip of beach on the marsh. As she approaches, the female may swim parallel to the shoreline, but when the tide nears its peak, she clambers up on the sand and moves away from the water's edge. Avoiding areas of dense vegetation, the terrapin begins searching for a nesting site on the top of the dunes or along its slopes. She probes the ground with her snout and, when satisfied, she pushes sand aside with her front feet. At some point the turtle moves forward slightly and continues to dig with her hind feet, creating a triangular-shape nest whose depth varies from 4 to 8 inches. Having finished her task, the female stands over the nest, bobs her head up and down and deposits four to 18 pinkish, leathery

43

Northern terrapin, Sherwood Island State Park, CT. Long Island Sound.

eggs. As each egg is laid, she covers it with sand. The female then caps the nest, pats it down and abandons her brood to the mercy of nature.

Protected only by a layer of sand, the eggs incubate over the next 60 to 104 days. During laying, gulls sometimes attempt to drive the turtle off while crows often dig them up as soon as the female has left the area. Together with foxes, raccoons and others, predators are known to destroy up to 75 percent of the nest.

The uppermost eggs in the nest are usually the first to hatch. Over the next one to nine days, the remaining members of the brood follow suit. Barely over an inch in length, the newborn terrapins claw their way to the surface and generally emerge during the daylight. They immediately strike out over the sand to the nearest vegetation where they remain hidden until nightfall. Some of the hatchlings, however, overwinter in the nest and emerge for the first time in the spring.

The newly hatched terrapins generally do not feed during the autumn of their first year. When the weather turns cold they dig into the mud and begin to hibernate. Adults at Cape May, NJ, often spend the winter buried in mud bottoms of tidal creeks. Though the turtles normally remain dormant throughout the cold season, they sometimes emerge from their refuge during unusually warm winter days.

As the days of spring grow longer, the young terrapins leave their shelter. The majority, however, continue to fast until at least seven to eight months of age. The turtles then begin to prey on mud crabs and a variety of other crustaceans. They also feed on blue mussels, salt-marsh snails, sandworms and fishes such as silversides. By the end of their second year, most have grown an additional inch in length and, over the next three years, the turtles gain two more inches. Though some females raised in captivity lay their first eggs at 4 years old, most do not reach sexual maturity until at least age 7. They produce their maximum number of eggs at about age 25. The life span of the terrapin is thought to exceed forty years.

Divided into seven subspecies, terrapins collectively range from Cape Cod, MA, to Texas. They inhabit coastal salt marshes, tidal creeks, brackish rivers, channels, estuaries and lagoons that are protected by a barrier beach. The turtles are regarded as "the only reptiles restricted to brackish water and saline tidal marshes in the United States" (Daiber, 1982).

> **Determining gender:**
> Females grow longer than males and their head is larger. Males are seldom longer than 5 inches, their nose is more pointed in shape and their tail is longer than that of the females.

During colonial days, terrapins were considered so common that their meat was frequently served to slaves. Slave owners and others, however, soon developed a taste for the reptiles. During the mid-1890s, the turtles were shipped to market from Buzzards Bay, MA, "by the barrel" and their population began to show signs of rapid decline. Believing that the species was doomed for extinction, the United States Fish Commission, in 1902, began investigating the possibility of raising them in captivity. In 1913, a company at Beaufort, NC, constructed an elaborate nursery and within a few years 15,000 to 20,000 terrapins were hatched annually. But the advent of World War I brought a significant increase in labor cost at the hatcheries, and, at the same time, demand for terrapins decreased. Though the loss of interest in the turtle as a food item helped preserve the creature, the human impact on its habitats continues to threaten it. ■

CREATURES OF THE SANDY BEACH AND SANDY BOTTOM

2

*W*hile examining a sample of intertidal sand under a light-powered microscope, it may be difficult to imagine the type of creature that thrives in this desert-like environment. The north and mid-Atlantic sand beaches and sand bottoms are, for the most part, composed of particles of quartz and feldspar that are constantly being lifted and carried by winds, waves and currents. Strong winter waves move the sediment off the beach and deposit it offshore as a sandbar. Gentle summer waves return the sand shoreward and rebuild the beach. Longshore currents slowly transport the sand parallel to the shoreline.

In this atmosphere of continual shifting and scouring, the substrate can neither provide for attachment of seaweeds and larger marine animals, nor would it seem capable of offering others food or a hiding place from predators. Yet after watching through the microscope for just a short time, it is apparent that there are a number of minute creatures moving through the sand. A thin piece of quartz flickers up and down as a roundworm (nematode) breaks its way to the surface. In another area, a tiny tube-shaped crustacean with a single red eye and short antennae emerges from the sediment.

46

The crustacean, a harpacticoid copepod, swims for a short distance and disappears as quickly as it made its entrance. These creatures are but two of a number of animals that have adapted to this harsh environment.

Harpacticoid copepod

The creatures of the sand and other marine environments are often divided into groups according to size. The macrofauna are generally regarded as animals that can easily be seen, while the microfauna are those that are invisible to the unaided eye. The intermediate size, meiofauna, are described as animals that are small enough to pass through a 0.5 mm mesh but are retained by a $5u$ mesh (u = micron = 1/1000 mm - 1 mm = .03937 in.). The meiofauna, also known as the interstitial fauna, can reach a length of 1-2 mm. They are, however,

Nematode worm

slim enough to slip through the spaces between the sand grains (interstitial space), usually without pushing the particles aside. An elongated, ribbon or tubular shape is characteristic of the meiofauna.

Harpacticoid copepods, nematodes, ciliates and flagellates, and other large protozoans are part of the meiofaunae. In addition, they include the especially strange-looking species such as the gastrotrich and tardigrade. The nematode, however, is the more common of the these creatures, representing about 70 percent of the meiofauna. On a Danish mud flat over 4 million nematodes were reported in an area of 1 square meter (10.76 ft²)! These creatures are found worldwide from the arctic to the tropics. They inhabit the land, fresh water and the sediments in the depths of the oceans. In the marine environment, certain nematodes feed on microscopic algae (diatoms). They also can attack larger organisms by using their "teeth" (cutting plates) and then sucking out the contents. Other species feed on bacteria, protozoa and detritus.

The smallest animals, the microfauna, are exclusively protozoans, one-celled creatures that belong to Kingdom Protista. The macrofauna, the largest of the creatures inhabiting the sands are the easiest to observe. A selected number of them are covered in this text. Like mud and rock, sand supports certain animals that live only in that environment. Others, however, are found in various mixtures of sand and mud or other substrates. Some live on the surface of the sand (epifauna) though they may occasionally bury themselves. Others live most of their lives in the sediment (infauna).

Shorebirds such as sandpipers, actively feed on the inverte-

47

brates living in the sediment. Scurrying along the edge of the receding tide, the birds probe the sand with their long slender beak. Then, as the next wave breaks over the holes, tiny creatures are washed to the surface and are carried seaward in the thin film of water; they are eagerly picked up by the foraging bird. ■

The invertebrates

Atlantic sand crab (=mole crab), *Emerita talpoida*

Habitat: Beach sand.

Other common names: Mole crab, sand bug.
Phylum: Arthropoda. Subphylum: Crustacea.
Class: Malacostraca. Order: Decapoda. Family: Hippidae.
Geographic range: Harwich, MA to Horn Island, MS.
Maximum size: Male 0.8 inches (19.4 mm).
Female 1.4 inches (36 mm).
Reproductive season: Winter to autumn depending on
location and temperature.
Egg color: Bright orange when first attached.
Dirty gray before hatching.
Life span: Females to 15 months, males 12 months.

*T*he Atlantic sand (mole) crab is a true creature of the sand. It is totally dependent on its chosen environment and, away from it, the animal cannot survive. During the warmer months of the year, the tiny egg-shaped crustacean lives at or below the surf line, preferring steeply sloped, wave-swept beaches. In the winter, it is also found in sand, but usually at depths of 9.8 - 13 feet (3 to 4 m).

In 1936, Lyman Jones of Southeastern Louisiana College noted that sand crabs were sometimes abundant on the beach, while at other times they were completely absent. Studying the movements of a closely related species, *Emerita emerita*, Jones tied a foot-long thread to one of the creatures' first legs.

Second antennae,
(plume-like)

Atlantic sand crab.
After Snodgrass, 1952.

48

Atlantic sand crab on a Connecticut beach, Sherwood Island State Park.

When the sand crabs buried themselves in the sediments, the thread stretched out on the surface in the wash of the waves. Using this method, he was able to follow tagged animals for up to a day and a half. He found that they were always in the wave wash, from the high to the low tide line.

The Atlantic sand crab also follows the tides. As the waters rise, the crab emerges from the sand on its own, or it is washed up by the breaking waves. It is then carried up the beach in the wave wash. When the power of the wash slackens, the creature burrows into the sand before it can be carried back down the beach. It positions itself at the edge of the surf, facing toward the sea. The crab digs with its tailfan appendages (uropods) and fourth pair of legs. Its first three pairs of legs operate somewhat like oars; they push the sand forward and propel the creature backward into the hole. Once buried to its eyestalks, the animal extends its plume-like antennae and begins to passively filter food from the receding waves. Using its mouthparts, the crab alternately removes trapped food from one antenna and then the other.

The sand crab responds immediately to a falling tide. It moves to the surface and it is carried back down the beach in the wave wash. As the force of the wash slackens, it reburies itself and begins feeding once more. The receding wave tends to move sideways (laterally) to the beach front. Since the crab is carried by the water, it is assumed that the creature moves in the same general direction. This may account for the large concentrations of sand crabs that are found at

49

certain sites along a beach. The crustacean's population along the cold New England shoreline is, however, much smaller than in the more southerly states; its breeding season also is apparently influenced by temperature and geographic location.

At the northern extent of its range, the sand crab's reproductive season begins in the late spring. Mating is thought to occur during a nuptial swim, and the male, who is often barely a year old, dies soon thereafter. As will be seen, however, this is not his first encounter with the opposite sex.

In another closely related species of sand crab, *Emerita analoga*, the female attaches her eggs to her specially adapted swimmerets (pleopods) some 12 hours after mating. The process probably takes about as long in the Atlantic sand crab. When the female first attaches her eggs, they are orange, but as they near hatching they become a translucent gray. Observed in the laboratory, the female releases clouds of larvae as she propels herself off the bottom in a series of short spurts. The tiny creatures swim as soon as they hatch, and they join other near-shore ocean drifters. Over an estimated three to four weeks, the planktonic larvae pass through some six to seven zoeal stages and metamorphose into their final, megalopa stage. At that point, the creatures very much resemble the adults. They easily burrow into the sand and they swim using their well-developed swimmerets. Unlike the larvae, adult males have no swimmerets and the females' are greatly modified. The adults of both sexes swim using their tail-fan appendages.

Larval Atlantic sand crab, zoea:
Except for a few red-colored spots, (chromatophores) the creature is transparent.
After Rees, 1959.

A newly settled male becomes sexually mature within a very short time of settling on the beach. Barely 0.1 inches (3 mm) long, he attaches himself to a female and deposits a ribbon-like package of sperm. He is then joined by as many as six more like-sized males, each of which mate with the same female.

By late summer or early autumn, the female has released her eggs and her tiny suitors have long since moved on; she mates only once and dies a short time after hatching her brood. The young males and females of that season, grow slowly until the next spring, and they may then grow as much as 0.08 mm in length per day. When mature, the females mate and the cycle is renewed. ∎

Sevenspine bay shrimp (=sand shrimp), *Crangon septemspinosa*

Habitat: Mainly sand and mud bottoms.

Other common names: Sand shrimp, salt and pepper shrimp.
Phylum: Arthropoda. Subphylum: Crustacea. Class: Malacostraca.
Order: Decapoda. Family: Crangonidae.
Geographical range: Gulf of St. Lawrence to eastern Florida.
Salinity tolerance: Wide range of salinity (euryhaline).
Oxygen requirements: 2-3 ppm marginal.
Maximum size: Length to approximately 2 3/4 inches (70 mm).
Reproductive season: In Delaware Bay, March to October.
Egg production: A 70 mm female can carry some 7,500 eggs.
Larval stages: Planktonic - approximately 8 stages.
Life span: To approximately 3 years.

*T*ypical of many creatures of the sand bottom, the sevenspine bay shrimp finds protection from predators by burying itself up to its eyestalks in the sediments. Its ash-gray color with a scattering of tiny dark specks over the back of its shell, helps it blend in with the sand. On a mud bottom, it assumes a darker color that serves as camouflage. The crustacean often remains inactive during much of the day. If threatened by an approaching predator, however, it bolts from hiding and swims a short distance to bury itself once more. As daylight fades to dark, the shrimp leaves the relative safety of the sea floor and assumes the role of a hunter.

Another crustacean, the mysid shrimp *Neomysis americana*, also abandons its near-bottom habitat after sunset. It is a sought-after food item for the sevenspine bay shrimp. Even in the absence of mysid shrimp, the sevenspine bay shrimp becomes more active under the cover of dark. But its movements, as observed in the laboratory, significantly increased in the presence of a potential meal. Though this nocturnal predator readily devours mysid shrimp, it also feeds on almost anything with which it comes into contact. In an aquarium, researchers have shown that even Baker's yeast or hard-boiled eggs stimulate a search by the shrimp. Following a zigzag search pattern, it homes in on the scent.

In nature, the shrimp's diet includes a variety of small crustaceans, nematode and polychaete worms, the tiny gem clam (*Gemma gemma*, see photograph page 53), diatoms and detritus. Its ability to obtain sustenance from a wide variety of sources has

51

Sevenspine bay shrimp hidden in the sand. Look for a pair of eyes in the photograph's lower right hand side. New Haven Harbor, CT.

The same sevenspine bay shrimp on the sand's surface.

allowed it to thrive and play a major role in the tidal marsh-estuarine food chain. The shrimp is a rich food source for many game fishes including the striped bass, bluefish, weakfish and summer flounder. In Long Island Sound, 50 percent of the fish

52

species depend on the crustacean for part of their diet.

The relative importance of sevenspine bay shrimp is obvious when it is realized that it is one of the most common bottom-dwelling (epibentic) invertebrates of southern New England. Over a 10-year period, it often comprised more than 90 percent of the invertebrates captured in trawl surveys of New Haven Harbor, CT. Though the 1 to 2-inch long crustacean is too small to be of commercial value, it was sold through New York markets during the late 1800s. Used for human consumption and as bait, one gallon of fresh shrimp fetched about $2. Its close European relative *Crangon crangon*, however, is today the basis of an important fishery. In Germany and the Netherlands, the annual yield of more than 5,000 tons is mainly processed into fish meal. Minor fisheries for the crustacean also exist in France, Belgium, Denmark, the United Kingdom, Spain, Italy and Morocco where some are used for human consumption.

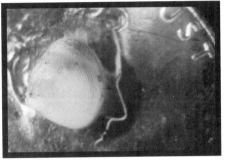

Gem clam on the head of a penny. 1/8 in.

The gem clam, *Gemma gemma*, is found in sand/mud substrates from Canada to Texas. It is an important part of the diet of many bottom feeders including the sevenspine bay shrimp. Sexes in the gem clam are separate. The female broods her young within her shell.

The sexes of the sevenspine bay shrimp can be easily distinguished by careful examination of the first pair of swimmerets (pleopods). In the male, the inner branch (endopod) of the first swimmeret is generally less than .04 inches (1 mm) long, while the female's can be .24 inches (6 mm) or more in length. The shrimp has at least two peak reproductive periods throughout its range. The first occurs during the spring and the next is in the mid- to late-summer and/or the fall. In the Mystic River estuary, CT, hatching seems to depend on temperature (50°F [10°C]), spring tides and sites where currents run parallel to the shore and out to the sea. The strategy helps disperse the larvae to deeper waters.

After passing through some eight larval stages, the shrimps settle to the bottom and begin to migrate back toward shore. The juveniles (under 0.8 inches [20 mm] in length) and first-year adults then tend to dominate the estuary's shoreline population. With further growth, the shrimps apparently lose their ability to tolerate low salinity and temperature extremes. They move offshore once more and return only to spawn.■

Longwrist hermit crab, *Pagurus longicarpus*

Habitat: Harbor beaches, harbor channels and along
shallow intertidal on a variety of bottoms.

Other common names: Long-clawed hermit crab, long-armed hermit crab,
hermit crab, dwarf hermit crab.
Phylum: Arthropoda. Subphylum: Crustacea. Class: Malacostraca.
Order: Decapoda. Family: Paguridae.
Geographic range: Nova Scotia to northern Florida.
Sanibel Island, FL, to Texas.
Depth range: Shallows to 200 m (656 ft).
Size: Male length .28 inches (7.25 mm), female length .18 inches (4.5 mm).
Color: Variable with locality.
Reproductive season: New England, May to mid-September.
Georgia, March to July.
Florida, September to April.
Texas, throughout the winter.
Egg production: 260-4,054 eggs.

*I*ts very existence is dependent on a borrowed home.
The tail or abdomen of the hermit crab does not have
a hard external skeleton, and without something to
protect it, the creature is fair game for any predator.
Thus, from the time its larva metamorphoses into a bottom-dweller,
the crab takes refuge in empty snail shells. As it grows, it must find
progressively larger shells. If the shelter is too small to allow the crab
to completely withdraw itself, it can easily fall victim to a predator.
When the creature is trapped intertidally, such a shell may also make
it more susceptible to desiccation. Cramped quarters can prevent the
animal from growing to its potential, and in some species, the female
cannot carry her normal number of fertilized eggs. Selection of the
right shell is then one of this crustacean's most important tasks.

The hermit crab generally finds a potential residence using
its compound eyes; the contrast of a shell with the background
apparently assists in locating it. When it comes into contact with the
shell, however, hairlike sensory setae on the crustacean's legs and
claws help it determine the shape of the shell. The crab rolls the shell
between its walking legs and examines the surface with its claws.
Once the opening has been located, the creature inserts one or both
of its claws, often reaching in as far as it can to explore the interior
surfaces. The hermit then pulls itself out of its old shelter, inserts its
abdomen into the new one and withdraws to the recesses of the shell.
In a snapping motion, the animal often stretches itself in and out of

the shelter several times before settling in. But if the new residence is not satisfactory, the crab reoccupies the old shell and continues its search.

The choice of a shell is based on shape, dimension, weight and covering. In some areas, however, the supply cannot keep up with the demand and the shell is then often less than ideal. In soft mud, the longwrist hermit crab has been observed occupying the shell of the solitary glass bubble, a snail whose shell is so delicate that it easily breaks with slight finger pressure. Alarmed by a prodding diver, a hermit crab usually withdraws to the safety of its borrowed home. It may also raise itself on its legs, push itself off with its claws and propel itself backward for a short distance. When occupying a glass bubble shell, however, the threatened creature often abandons the fragile shelter.

On beaches subject to strong wave action, many of the empty snail shells are damaged. A longwrist hermit crab occupying a damaged shelter is generally more aggressive, and fights may ensue over a neighbor's shell. When a battle develops, the attacker reaches into the defender's shell with its claw(s) and maneuvers the shell's opening to a position directly opposite its own. The attacker then holds the opponent's shell with its first two walking legs and forces the shells together in a series of

Switching shells.

Top: Longwrist hermit crab explores a new shell with its claws.

Middle: Pulling itself out of the old.

Bottom: Occupying the new shell. If it is not satisfactory, the hermit crab returns to its old shell.

rapid blows. The action generally results in eviction of the defender but without doing physical harm to either party. The larger crab is the usual winner.

Many hermit crabs occupy a specific species of shell. In Long

Longwrist hermit crab with snail's fur on its shell, Norwalk, CT.

Island Sound, the juvenile and female longwrist hermit crabs tend to prefer those of the eastern mud snail and the Atlantic oyster drill. The adult males, which are always larger than females, normally occupy the common periwinkle's shell. But these crustaceans are not always the sole occupants of their mobile homes. They often share their shell with other creatures.

The tiny polychaete worm, *Polydora commensalis*, is apparently found only in shells occupied by hermit crabs (see page 57). The worm builds itself a tube by boring a deep groove in the center of the shell (columella) and covering it with a thin calcareous layer. Often, both the male and female polychaetes are found in the same shell with their tubes lying alongside each other. During feeding and presumably during mating, the worms extend themselves partially out of their tubes. The eggs are retained in the tube, and upon hatching the larvae are released into the plankton. The precise relation between the worm and its host is not fully understood.

The surfaces of some shells are occupied by barnacles, slipper shells or a hydroid known as snail fur (*Hydractinia spp.*). The longwrist hermit crab generally avoids a shell inhabited by barnacles and accepts the presence of a slipper snail, as long as it does not make the shell too heavy. A shell covered by *Hydractinia*, however, is actively pursued. The hydroid's stinging cells apparently protect the crab from certain predators.

The longwrist hermit crab is also a predator, but the activity

56

accounts for only a very small part of its diet. Most of what it eats is scavenged (45 percent) and nearly as much nutrition is obtained from detritus and materials associated with sand grains (40 percent). When feeding on a sandy bottom, the crab scoops up bits of substrate with its claws, and sorts out food with its

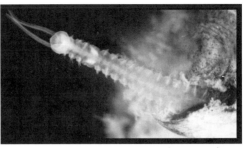

The worm *Polydora commensalis* occupying a longwrist hermit crab's shell.

mouthparts. Scavenged morsels are grasped and torn by the claws, and carried to the mouth. The crab's use of surface foam as a food source, however, is one of its most fascinating behaviors. Positioning itself in water less than an inch deep, the creature rolls itself over on the back of its shell and extends its legs above the surface. It then sweeps the foam to its mouthparts and removes any edible matter.

The smell of food attracts a hermit crab sooner than any visual stimulus. When scavenger material is detected, a large number of the crabs generally gather, and, because of their numbers, they probably keep other species away from the feast. In a behavior reminiscent of cattle egret following livestock for an opportune insect, large numbers of hermit crabs have also been observed closely following horseshoe crabs that are plowing into the sand. The hermits apparently feed on tiny organisms kicked-up in the sediment.

Along the coast of New England, the longwrist hermit crab's reproductive season begins in early May and continues through mid-September. In a closely related European species, *Pagurus bernhardus*, the onset of breeding season finds a male approaching a receptive female from behind. Taking hold of rim of her shell with his small claw, he keeps his larger claw free to ward off other suitors. The female is then maneuvered so they are face-to-face. Using his walking legs to hold his mate's shell steady, the male alternately attempts to coax and pull her partially out of her shell. In the encounter, there is often a mutual touching of claws and eventually both

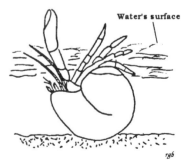

Longwrist hermit crab. Using surface foam as a food source in approximately 25/32 in. (20 cm) of water.

Redrawn after Scully, 1978.

57

extend their bodies nearly out of their shells; walking legs are intertwined and mating takes place. The longwrist hermit crab's ritual is very similar.

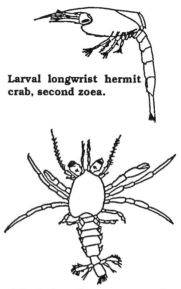

Larval longwrist hermit crab, second zoea.

Sometime after mating, the longwrist female attaches her 300 to 4,000 eggs to the tiny appendages (pleopods) on her abdomen. She carries and protects her brood within her shell. As the eggs begin to hatch, she releases her larvae into the plankton by moving her abdomen in and out of the shell. Over some three weeks, the larvae pass through three mosquito-like zoeal stages and a crab-like megalopa stage. They then metamorphose into bottom-dwelling crustaceans (glaucothoe).

Final larval stage, megalopa.
Redrawn after Roberts, 1970.

The newly settled hermits take up residence in the shallows, on a variety of bottoms, and they immediately begin searching for an appropriate sized shell. Growth is rapid, and they can reach maturity in about four months. Within a year, most have reached their maximum size. When the crabs make their home on a loose sandy bottom, they often bury themselves in the sediments with only their eye stalks left uncovered. Unless disturbed, they often remain there during most of the day and emerge at twilight. In the fall, as the coastal New England water temperatures approach 50°F (10°C), the creatures begin their migration to deeper water. By the time these temperatures have dropped to 39°F (4°C), offshore movement ceases. The crabs scoop out a depression in a muddy bottom and the settling sediments tend to cover them. The crustaceans then remain there for the winter.

Spring warming brings a renewal of activity and the first to arrive along the shoreline are juveniles and young adults. The larger, older crabs tend to migrate farther from shore, and they are usually the last to return. ∎

Horseshoe crab, *Limulus polyphemus*

Habitat: Estuaries to continental shelf.

Other common names: King crab, swordtail crab, pan crab,
                    horsefoot crab, horsefeet crab, blue-blood.
Phylum: Arthropoda. Subphylum: Chelicerata.
Class: Merostomata. Family: Limulidae.
Other species: Found in the Indo-Pacific, *Tachypleus tridentatus*,
              T. gigas, and *Carcinoscorpius rotundicauda*.
Geographic range: Maine to the Yucatan.
Reproductive season: Late spring to early summer -
                    depending on the area.
Egg production: Maximum of about 88,000 per season.

*T*he name horseshoe crab is misleading. The creature was once considered part of the crab family, but after having studied the animal in 1829, zoologist Stauss-Durkheim suggested that it was in actuality related to arachnids. His findings became widely accepted in 1881, following the publication of *Limulus an Arachnid* by Ray Lancaster. Placed in the arthropod subphylum of *Chelicerata*, this ocean-dwelling creature shares a close relationship with scorpions, spiders, mites and ticks.

The horseshoe crab predates the time when giant fish-lizards - *Ichythyosaurus* - roamed the seas and dinosaurs ruled the earth. A true living fossil, it traces its ancestry to the middle of the Cambrian period, some 500 million years ago. Only four species of horseshoe crabs roam the seas today. Three of them are found in the coastal seas of Asia and the fourth, *Limulus polyphemus*, inhabits the North American coastline from Maine to the Yucatan. Each spring, throughout its range, the warming coastal waters greet the returning *Limulus* from its deeper-water, winter refuge. As it has for millions of years, the migration marks the beginning of a new life cycle.

Precisely what initiates the animal's trek from offshore waters and how it finds its way is not yet known. When it arrives at the shoreline, *Limulus* generally chooses to spawn on a beach that is well protected from heavy surf; it apparently can then orient itself to the beach front in response to the wave

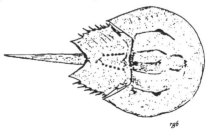

rgb

59

surge. Timing its reproductive activity to coincide with the high tide of a new or full moon, the male joins others of his sex in patrolling the shoreline. When the suitors detect a female that is ready to spawn, they jostle to position themselves behind her. The lucky one takes hold of her carapace with his specially adapted claw and clings to her as they move in tandem to near the edge of the high tide line. Once there, the female digs a shallow nest, deposits some 3,000 greenish eggs and the male releases his milt. Still joined, the pair then retreats to below the low tide line, often remaining together to return to spawn on another high tide. During the season, a female can produce an estimated 88,000 eggs.

The nuptial ritual is not usually performed in privacy. Other males, as many as 10, follow the pair up the beach and some are believed to also deposit their milt in hope of fertilizing a few of the eggs. The first external signs of successful fertilization appear within 10-15 minutes of spawning. Tiny pits form on the egg's surface. Over the next few hours, the egg's surface becomes smooth, then granulates and becomes smooth again. Development proceeds rapidly.

By the fifth day the embryo has grown rudimentary appendages and some nine to 10 days later, in synchrony with the next spring tide, the creature hatches. In the northern Gulf of Mexico the eggs hatch after about five weeks. At sites where the nests are subtidal, the process may take longer.

The rising waters of a spring

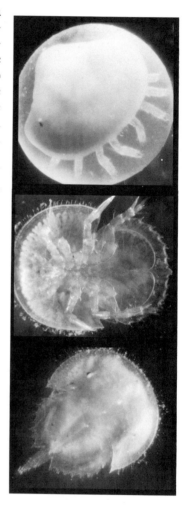

Developing horseshoe crab

Top: Developing embryo, fourth molt. Before hatching, the embryo molts four times within its egg.

Middle: Newly hatched, 3mm long larva as it is found in the plankton. Note the lack of a tail.

Bottom: First-tailed, bottom-dwelling juvenile.

Small pincers (chelicerae): The crab's first pair of appendages are small pincers used in picking up and crushing food.

Claspers: The male's fist pair of walking legs have a modified claw (clasper) at their tip; the clasper somewhat resembles a boxing glove.

Claws: The female's first pair of walking legs have tiny claws at their tip.

Legs: The horseshoe crab has five pairs of legs. The female's first pair of legs ends with a small claw, while the male's have hooklike claws (claspers). They are used in clinging to the rear of the female's shell during spawning. For both sexes, the next three pair of legs pincer at their tips and the last pair end in leaflike appendages used for pushing and clearing sediments during burrowing.

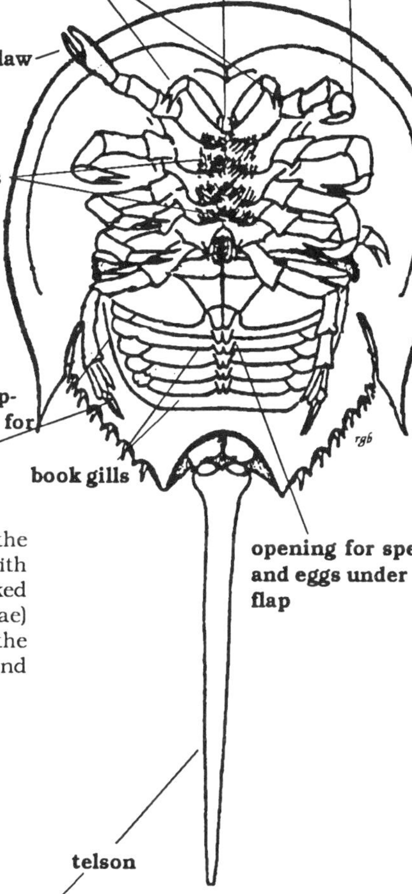

small chelicerae (pincers) male clasper

mouth

female claw

leg spines

leaflike appendages for digging

book gills

opening for sperm and eggs under the flap

Leg spines -gnathobases: The base of the first four pairs of walking legs are lines with spines called gnathobases. Food is picked up with the crab's tiny pincers (chelicerae) and is passed to the leg spines. As the animal walks, the food is broken up and moved toward the mouth.

Gills -book gills:
At the rear of the animal, on the underside of the first flaplike structure, are located openings for the male's sperm and the female's eggs. The underside of the remaining flaps consist of leaflike folds known as book gills.

telson

Tail -telson: The horseshoe crab's tail is not a weapon; it is used by the animal to upright itself when it is upside down.

Horseshoe crab, *Limulus polyphemus,* **ventral (underside).**

tide free the larva from its sand nest and carries it down the beach. Resembling a trilobite (see photograph, page 60), the 0.12 inches (3 mm) long, tailless animal retains some yolk and, during this stage, it is assumed to require little or no food. Using its gills to propel itself, the animal swims in an upside-down position; the activity may help to disperse the larval population. After some six to 21 days, the trilobite larva sheds its external skeleton (molts) and acquires a telson (tail). The creature, known as a first-tailed juvenile, then spends most of its time on the bottom where it starts imitating the burrowing behavior of an adult. It also begins to

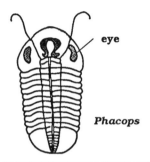

Phacops

The trilobite *Phacops*. Over 3900 species of these primitive arthropods roamed the prehistoric seas. They became extinct during the late Paleozoic period.

feed. In captivity the creature accepts anything from bits of blue mussel to algae; in nature it feeds on detritus.

Most of the newly hatched horseshoe crabs are believed to spend their winter as first-tailed juveniles buried in the sediments of the intertidal zone. The male reaches sexual maturity after some 16 plus sheds, at about 9 years of age. The female matures following approximately 17 molts, around the age of 10. In the weeks preceding shedding, a soft, wrinkled skin forms on the underside of the shell. When the time is right, the creature burrows into the sediments, below the low tide line. The front of its carapace then begins to split, and the creature emerges from its old shell. In the process the horseshoe crab's body absorbs sea water and increases in size by as much as 25 percent.

As *Limulus* molts and grows, it moves farther offshore and sometimes migrates as far out as the continental shelf. Once mature, it begins its march back toward the shoreline. The ungainly-appearing creature can move fairly rapidly across the sea floor. Like its trilobite larva, the younger animal can also swim. To launch itself off the bottom, *Limulus* sometimes climbs the side of a steep rock and propels itself with coordinated movements of its walking legs and gills. It can also launch itself from an upsidedown position on a sand substrate. Swimming upsidedown, it travels for just a short distance before sinking to the sea floor. The creature generally lands in an upsidedown position and then begins to turn itself over by arching its body and digging its tail into the sediment. When sufficiently tipped to one side, the animal takes hold of the bottom with its legs and flips itself upright.

The horseshoe crab is also very adept at burrowing. It first digs a shallow hole in the sediments and arches its body so that the front edge of its carapace slides down into the depression. Pushing itself farther into the bottom, the animal plows its way until at least the front part of its shell is completely covered.

Limulus can find prey that is hidden in the sediments. It feeds on various worms and bivalves such as the minute gem clam, *Gemma gemma*, and the commercially important soft-shelled clam, *Mya arenaria*. Digging into the bottom, the animal grasps a shelled-victim with its small pincers (chelicerae), and passes it to a series of hard spines (gnathobases) that line the base of its walking legs. The spines then grind up the material as it moves toward the mouth. Once swallowed, the food is further ground up and sorted in the creature's gizzard and any undigestible matter is regurgitated. As it is with other coastline creatures, *Limulus* the predator is also prey.

The horseshoe crab's strategy of laying its eggs near the high tide line helps protect them from water-dwelling predators. When the nest is left uncovered by the receding waters, however, the brood becomes easy pickings for many seabirds. Adults that are stranded intertidally may also fall victim to seabirds. Below water, certain species of fishes and sharks feed on the trilobite larvae, juveniles and/or adults. Though not ordinarily consumed by humans, *Limulus* has proven to be of great benefit to man.

The Native Americans of Roanoke Island, NC, are said to have used the horseshoe crab's tail as a spear-tip for fishing, and the animal is still used as bait for eels. For years, *Limulus* was spread over fields as fertilizer, but the practice has nearly ceased. More recently, it has gained importance in the research community. The creature's compound eyes have become an investigative tool and its blood has come into use in the quest to conquer human disease.

Located on the sides of the animal's shell, the compound eyes are made up of eight to 14 individual units (ommatidium), each of which houses a lens and cornea. The eyes probably detect movement but they are apparently unable to form an image. Their simplicity, however, and the size and accessibility of their optic nerve, and their ability to produce an electrical

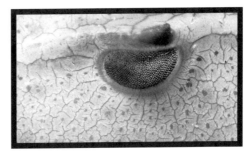

The compound eye of *Limulus*.

63

Female *Limulus* burying her eggs with a male clinging to her shell. Note the male at the right using his telson to turn himself over. Cape May, NJ.

response similar to that of more complex eyes, have made them exceptionally valuable. In 1967, H. Keffer Hartline of Woods Hole, MA, was awarded the Nobel Prize for his work on the horseshoe crab's optic nerve and vision.

Limulus blood is cloudy white in color, but when it comes into contact with air it changes to a cobalt blue; hence the animal's name blue-blood. The blood's only circulating cell, the amebocyte, contains all of the elements of the horseshoe crab's clotting system. When the creature begins to bleed, the amebocytes flatten themselves and develop long processes that reach out from cell to cell, eventually blocking further loss of blood. While studying this mechanism during the 1950s, Woods Hole scientist Frederick Bang discovered that the blood also clotted in the presence of certain poisons produced by bacteria (endotoxins). From this work emerged a medically important test for bacterial contamination and a diagnostic tool.

To obtain *Limulus* blood, a large gauge needle is inserted into the sac surrounding the creature's heart. Some researchers feel that as much as 30 percent of the total blood volume can be drawn without harming the animal. A substance, *Limulus* amebocyte lysate (LAL), is then extracted from the blood. When LAL - a freeze-dried powder - is added to a test solution, it can detect minute amounts of endotoxin. It is used in checking a variety of pharmaceuticals and health-care delivery products such as kidney dialysis machines, syringes and catheters. It has also been investigated for use in the diagnosis of meningitis, blood poisoning, gonorrhea and urinary tract infections.■

Sandworm, *Nereis virens*

Habitat: Sandy muds to fine sand; prefers muddier substrate.

Other common names: Clamworm, ragworm.
Phylum: Annelida. Class: Polychaeta.
Order: Phyllodocida. Family: Nereididae.
Geographical range: Both sides of the North Atlantic. In Europe, south to France, and south to Virginia on the US coast.
Salinity tolerance: Wide range of salinity (euryhaline), as low as 0.5 ppt.
Temperature tolerance: To as high as 99.5°F (37.5°C).
Maximum length: Approximately 27.5 to 29.5 inches (70 to 75 cm).
Color: Male iridescent blue - green. Female dull green.
Reproductive season: Full moon and temperatures above 45-46.4°F (7-8°C).
Egg production: 50,000 to 1.3 million depending on the size of the female.
Planktonic period: Trochophore larva -15 hours.
Life span: Approximately 3 years. At the colder part of its range, approximately 5 to 6 years.

*F*ace to face, the sandworm is an awesome sight. It has four eyes and four antennae on the front of its head, and four pairs of feelers (tentacular cirri) rising from the sides. Its proboscis - essentially its throat - houses two black-colored, zinc-hardened jaws that are armed with five to 10 teeth. Also equipped with a well-developed sense of smell, it is an efficient predator.

The worm lies in wait in its mucus-lined burrow, ready to pounce on any prey in reach. Turning its proboscis inside out, the predator snatches up the victim in jaws powerful enough to crush a small clam. The sandworm consumes some plant material but feeds mainly on small marine animals. In one study, the sandworm

zinc-hardened jaws

proboscis

antennae

tentacular cirri
(see text).

Sandworm.
From Wilson, 1988

Face-to-face view of a sandworm.

Long Island Sound.

65

significantly reduced the population of a tiny tube-dwelling amphipod crustacean, *Corophium volutator*. But its role as a predator is not one-sided. The bloodworm is known to prey on the sandworm, and the latter serves as food for a variety of fishes and certain shorebirds. The sandworm is also commonly harvested as bait, and researchers use it to study heavy metals and organic pollutants.

Amphipod
***Corophium* spp.**

The sandworm inhabits sandy muds and fine sands, producing a series of interconnected burrows that lie in the upper 25/64 inches (10 cm) of the muddier substrates. Though it seldom leaves its shelter except for spawning, it has been observed swimming and being carried by tidal currents. The non-reproductive swimming behavior is believed to help it find a more suitable habitat (see Bloodworm, page 14).

Spring water temperatures in excess of 45°F-46°F (7°C to 8°C) and the lunar cycle marks the beginning of the worm's spawning period. In preparation, the male's body undergoes dramatic change into its reproductive form known as epitoke. Then, during a lunar spring tide, the male rises to the water's surface and joins hundreds of other male epitokes in a spawning swim. Moving in a tight circle or in somewhat of a straight line with occasional tumbling and back-swimming, the male releases sperm and dies. The female does not make the change into an epitoke and remains at her burrows where she releases her eggs; she also dies shortly after spawning. Fertilization occurs by chance.

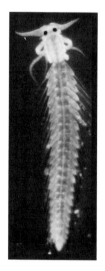

Epitoke, Long Island Sound.

The hatched larvae resemble a ball with a girdle of long hairs (cilia) surrounding their middle. At first, they barely move, but with increasing maturity they can intermittently lift themselves a few millimeters off the bottom. By the sixth day, the larvae are well into their trochophore stage and they join the plankton for some 12 to 15 hours. Their brief interlude as drifters is believed to allow for their limited distribution and, at the same time, minimize their movement into areas

Trochophore larva

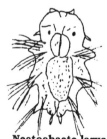

Nectochaete larva
After Bass, 1972

66

that might be unfavorable to further development. Around the seventh day, the larvae change into their nectochaete stage and they swim near the bottom in a corkscrew fashion. Some five days later, they cease swimming and begin feeding on the sandy bottom. At the age 2 to 3 years, the sandworms become sexually mature (5 to 6 years in the Bay of Fundy). Then, when conditions are right, they spawn and die. ■

The fishes

Northern searobin, *Prionotus carolinus*
Striped searobin, *Prionotus evolans*

Habitat:
P. carolinas - shows preference for smooth sand
but also found on rock and mud bottoms.
P. evolans - Sand, mud and rock bottoms.

Other common names:
     P. carolinas: Common searobin, green eyes, Carolina searobin.
     P. evolans: Southern striped searobin, red-winged searobin, flying fish.
Phylum: Chordata. Class: Osteichthyes. Order: Scorpaeniformes. Family: Triglidae.
Geographical range:
     P. carolinas: Bay of Fundy to eastern Florida.
     P. evolans: Nova Scotia to the east coast of Florida.
Depth range:
     P. carolinas: From 16 to 560 feet, (9 to 170 m).
     P. evolans: From 16 to 480 feet, (9 to 146 m).
Maximum length:
     P. carolinas: 15.75 inches (400 mm) - seldom over 12 inches.
     P. evolans: 18 inches (460 mm). Females mature at 5.5 in. (140 mm).
Salinity tolerance:
     P. carolinas: 5 to 32.2 ppt
     P. evolans: 7 to 32.8 ppt
Reproductive season: May through early August for both species in
                 southern New England.

A sandy bottom easily conceals the hunter and the hunted. The northern searobin, the predator, hides in the soft sand with little more than the top of its head and its emerald-colored eyes exposed. From this position, the fish waits for passing shrimp or another suitable meal; the sands also afford it protection from larger hunters. But the sudden approach of a SCUBA diver sends the searobin fleeing its shelter. As it swims rapidly away, the creature's large pectoral (side) fins trail close to its body. In slowing, however, the searobin spreads its pectoral fins outward like a pair of oval wings and glides to the

67

bottom. Coming to rest, it holds itself up on its tail and a pair of three finger-like feelers. The feelers (fin rays), are located forward of the wing-like pectorals. Part of the pectoral fins, they are characteristic of searobins.

The searobin "walks" the sea floor on its feelers as it searches for food. Though the appendages have neither taste buds nor receptors for smell, they

Striped searobin, gliding on its large pectoral fins, Newport, RI.

have nerve endings that can detect low levels of certain chemicals. Using this chemical sense, the fish can locate a concealed prey. To reach such a creature, it often digs into the sediments or stirs up seaweeds with its feelers.

The fish is a voracious feeder. It greedily ingests almost anything within its reach. Seaweeds, insects and even plastic balls have been found in its gut. Young northern and striped searobins, however, tend to consume mostly small crustaceans such as copepods. The growing youngsters gradually select larger prey including sevenspine bay shrimps (=sand shrimp), mysid shrimps, cumaceans, mud shrimps (*Upogebia affinis*), lobsters, and a variety of small crabs, fishes, mollusks and worms.

Searobins are well-known to fishermen as bait stealers; they readily take bait both at the bottom and at the surface. When hooked or while being handled, they invariably make loud clucking or grunting sounds. As a group, they are reputed to be the noisiest fishes on the New England shoreline. The coastal Atlantic's oyster toadfish is, however, capable of generating the loudest sound; it produces a grunt that can exceed 100 decibels (see Oyster toadfish, page 125).

Sounds made by searobins may at times be an alarm signal, but many others seem to be produced spontaneously. Some loud barks are thought to be in response to sounds generated by other species such as the oyster toadfish. Fisheries researcher Marie Fish suggest the barks are probably territorial in nature.

The northern and striped searobins produce their sounds with their balloon-like air bladders. Consisting of two lobes, these organs take up more than half the body cavity. When a bladder's muscle fibers contract, the entire organ becomes taut and the

68

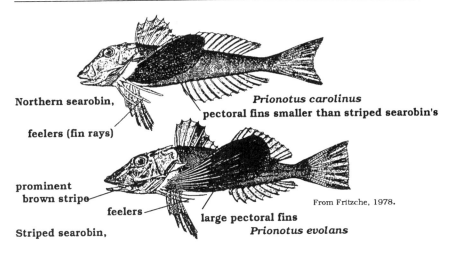

Northern searobin,   *Prionotus carolinus*
pectoral fins smaller than striped searobin's

feelers (fin rays)

prominent
brown stripe

feelers   From Fritzche, 1978.

Striped searobin,   large pectoral fins
Prionotus evolans

trapped gases are set in vibration. Both sexes produce sounds, and it is apparently during the spawning season that these fishes are the most vocal.

The months of April and May see the return of the northern and striped searobins to southern New England's shoreline; they spend the winter in deeper offshore waters. Both species begin spawning as early as May and continue into the beginning of August. Their eggs are buoyant and are carried by the currents. Depending on temperature, the northern searobins' eggs hatch some 60 to 90 hours after spawning. By the end of their first summer in Long Island Sound, their young have grown to a length of 1.5 to 3 inches (40 to 80 mm). At 1 year they are

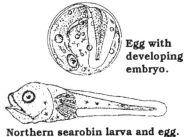

Egg with developing embryo.

Northern searobin larva and egg.

generally 5.5 inches (140 mm) in length and at 2, they are at least 9 inches (230 mm) long. Their maximum length is about 15 3/4 inches (400 mm).

Along the Atlantic coast, the Gulf of Mexico and the Caribbean, there are 15 species of *Prionotus* and four of *Bellator*; all belong to the family of Triglidae known as searobins. These fishes all have hard, bony heads armed with spines. Though they superficially resemble members of the sculpin family, they can be easily distinguished; only the searobins have the pair of three finger-like feelers.

Winter flounder,
Pleuronectes (=Pseudopleuronectes)
americanus.

Habitat: Juveniles, sand, sand/mud.
Adults, mud, sand, gravel.

Other common names: Blackback, Georges Bank flounder, rough flounder,
flounder, flatfish, lemon sole, sole, mud dab.
Phylum: Chordata. Class: Osteichthyes.
Order: Pleuronectiformes. Family: Pleuronectidae (righteye flounders).
Geographic range: Northern Labrador to Georgia.
Tolerances: Avoids water colder than 32°F (0°C). Prolonged exposure to
salinity below 15 ppt may be lethal.
Size: Reported up to 25.2 inches (64 cm) long. Larger fish from Georges Bank.
Reproductive season: January to May in New England.
Temperature at spawning: 34-50°F (1-10°C), peak at 35.6-41°F (2-5°C).
Spawning time: Between 10:00 pm and 3:30 am.
Egg size: Demersal (sink to the bottom) and round, .71-.96 mm.
Egg production: A female can produce in excess of 3 million eggs.
Metamorphosis: At about 2 months old.
Life span: maximum may exceed 12 years.

*T*he winter flounder begins life resembling most other
fishes; it has eyes on both sides of its head. Measuring
barely 0.1 inches (3 mm) in length, the newly hatched
larval fish is a feeble swimmer. It can propel itself off
the bottom but it immediately sinks back down. Though the larva's
ability to swim improves as it grows, it continues to alternate between
resting on the sea floor and swimming with the plankton. The
behavior apparently helps the creature survive by
allowing it to remain near the spawning grounds. If
swept out to open water, it is believed to have little
chance of finding sufficient food or avoiding predators. But in its shoreline habitat, the larva is still
subject to predation. Its main enemy is the tiny
medusa stage of the hydroid, *Sarsia tubulosa.* Using
stinging-cell-armed tentacles, *Sarsia* traps and consumes large numbers of the larval fishes; prey and
predator are abundant at about the same time during
the spring.

The larval flounder feeds on microscopic algae and invertebrates. As the fish approaches

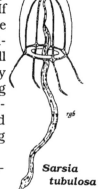

**Sarsia
tubulosa**

70

Larva 19 days old, 0.17 inches (4.2 mm) total length (TL).
Eyes on both sides of the head.

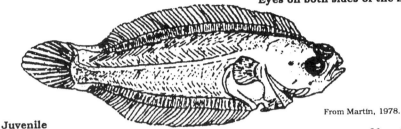

From Martin, 1978.

Juvenile
.31 inches (8 mm) TL. Metamorphosis is complete at 6.5 to 9 mm of length.

1/4 inch (6 mm) in length, its left eye begins to migrate toward the right side. The physical differences with other species of fishes then become apparent. Metamorphosis proceeds rapidly. Color on the flounder's left side gradually fades to white and the left eye moves totally to the right side. The change is complete by the time the fish is .31-.35 inches (8-9 mm) long, and the flounder then swims and lies on the bottom with its blind or left side down.

The young fish settles near shore, mainly on a sand or sand/silt bottom. It remains in the vicinity for the first two years of life and only migrates to deeper water to avoid temperature extremes. The adult flounder tends to prefer mud or eelgrass bottoms, but it also takes up residence on sand and even gravel and cobblestone. In mud and sand, the fish flips up the sediments and works its way into the bottom, often leaving only its eyes exposed.

The flounder feeds only during the daylight hours. Arching itself on its fins with its head raised off the sea floor, the fish's turret-mounted eyes scan the immediate area; its eyes move independently of each other giving it a better view of the bottom. When the flounder discovers a potential meal, it turns itself toward the creature and ceases breathing. Then, springing off the bottom, it snatches up its quarry. If a prey is not sighted, the flounder moves to a new location, usually no more than 3 feet away. During feeding it sometimes changes location as many as four or five times a minute. At night, the flounder lies still on the bottom with its eye turrets retracted.

The adult is an opportunistic feeder. Its diet includes a variety of small crustaceans, worms, mollusks, hydroids, sea stars, sea cucumbers, fish fry, fish eggs, insect larvae and even algae. During the cold of winter, the flounder apparently does not require food. It fasts from about November to April.

71

Winter flounder. The male's blind side is rough while the female's tends to be smooth. Large females, however, often have rough-scaled blind sides.

The adult's chosen habitat is largely regulated by water temperature. During the summer, the adult migrates to deeper waters to avoid temperatures above 59°F (15°C). It returns to the shoreline as the waters cool in early autumn, often reoccupying the same general area year after year. The winter flounder is well equipped for surviving the cold, inshore waters; it spawns in the shallows during the coldest part of the season. Within its tissues, the creature produces an organic antifreeze compound that lowers the freezing point of its body fluids.

In New England, the spawning season begins in January and continues through May. The height of activity in the Woods Hole region occurs in waters that are 32°F to 35°F (0°C - 1.6°C). The female begins her spawning run by propelling herself in a counterclockwise circle that is about a foot in diameter. As she swims upward, she releases her eggs and spreads them over a wide area. The male then similarly releases his milt. The eggs sink to the bottom and stick together in clumps; they hatch in about 16-24 days. ■

The winter flounder is a right-eye flounder. Its left eye migrates to its right side and the fish then lies down on its left side.
The summer flounder (page 73) is a left-eye flounder. Its right eye migrates to its left side and the fish then lies down on its right side.

Summer flounder, *Paralichthys dentatus*

Habitat: Sand and mud bottoms.
Estuaries during the summer, offshore during the winter.

Other common names: Fluke, common flounder, doormat flounder, plaice, chicken halibut, brail, turbot, flatfish, long-toothed flounder.
Phylum: Chordata. Class: Osteichthyes.
Order: Pleuronectiformes. Family: Bothidae (lefteye flounders).
Geographic range: Nova Scotia to south Florida. Greatest population between Cape Cod, MA, and Cape Hatteras, NC.
Tolerance: Withstands low oxygen. Temperatures from 6.6 -31.2°C (44-88°F), salinity nearly fresh to 37 ppt.
Size: Record - 1975, Montauk, NY, 22 pounds, 7 ounces.
Reproductive season: North of Chesapeake Bay, September to December. South of Chesapeake Bay, November to February.
Egg production: Ranges from 463,000 to over 4 million.
Hatching: 2-9 days after spawn.

*T*he summer flounder employs an unusual strategy to assure reproductive success. After having spent the warmer months in coastal waters, the adult migrates offshore to the continental shelf where it prepares to spawn. North of Chesapeake Bay, the breeding season generally begins in September; south of the bay, it begins in November. During the next few months, the flounder moves along the shelf gradually releasing up to 4 million eggs. By dispersing its eggs over a wide area, the fish presumably avoids large concentrations of eggs and larvae. The behavior is believed to help reduce predation and the impact of adverse environmental conditions on its developing offsprings.

Carried by the currents, the fertilized eggs incubate for some 74 to 94 hours. The larval fishes are less than 0.1 inches (3 mm) long as they emerge from

Fertilized egg, 0.8 mm

Larva, 75 hours after fertilization, 2.78 mm.

Larva, fin rays developing, 9.47 mm.

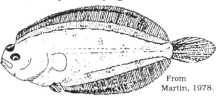

From Martin, 1978.

Larva, 10.4 mm. Right eye moving to the left side.

73

their encapsulated environment. They have eyes on both sides of their head and carry yolk on their underside. Within 96 hours of hatching, the larvae have exhausted their supply of yolk and must then feed on microscopic animal plankton (zooplankton). Once they have grown to a length of .35 to .47 inches (9 to 12 mm), their right eye migrates toward the left side. Metamorphosis is complete at approximately 19/32 inches (15 mm) and the creatures lie on the bottom with their right side down (See Winter flounder, page 70).

In North Carolina, larval and metamorphosing fishes enter the estuaries around February; in New Jersey, they first appear in October. Providing that the water temperature does not drop below 35.6°F (2°C), many of the young-of-the-year survive and flourish in southern New Jersey marshes. Taking advantage of the rising tide, the growing youngsters reportedly move up the marsh creeks to feed on the large concentrations of sevenspine bay shrimps, Atlantic silversides, mummichogs and marsh grass shrimps. As the tide begins to ebb, the flounders move back out to the bay. They grow rapidly during their first season in the estuary, reaching a length of 6.3 to 12.6 inches (160 to 320 mm) by summer's end. In the fall, they join the adult migration deeper water and remain there until spring. The young-of-the-year reach sexual maturity at 2 to 3 years old.

The adult flounder is a ravenous predator. Its large mouth allows it to easily swallow a 1-year-old winter flounder, small menhaden, squid, blue crab, hermit crab and a variety of other creatures. Lying motionless and partially buried in the sand, it waits for a passing prey. If one approaches too closely, the alert fish springs from hiding and snatches its victim. The summer flounder, however, is not restricted to bottom feeding. It also actively pursues baitfish at the surface. In the heat of the chase, it has been observed jumping clear of the water. Because of its size and feeding habits, the adult is generally regarded as being at or near the top of the estuarine food chain.

Most summer flounder taken by anglers weigh between one and three pounds. One that tips the scale at five pounds is considered a great catch. A summer flounder that is six or more pounds is referred to as a doormat flounder. Though a fish over 15 pounds is relatively rare, one taken off Fishers Island around 1915 weighed in at 30 pounds. More recently, in 1975, a 22 pound, 7 ounce doormat was landed on rod and reel at Montauk, NY. ■

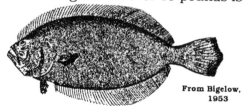

From Bigelow, 1953

Summer flounder, adult.

74

Bluefish, *Pomatomus saltatrix*

Habitat:
Adults pelagic (open ocean), estuaries, harbors
and off sandy beaches.
Juveniles over sand/gravel bottoms in the
Delaware River.

Other common names: Blue, snapper blue, chopper, tailor,
elf, fatback, snap mackerel, skipjack,
skip makerel, horse mackerel, greenfish.
Phylum: Chordata. Class: Osteichthyes.
Order: Perciformes. Family: Pomatomidae.
Range: Nova Scotia to the tip of Florida. Gulf of Mexico
to northern Mexico.
Salinity tolerance: Down to approximately 7 ppt.
Temperature tolerance: To 58°F (14.5°C), possibly to 50°F (10°C).
Maximum adult size: Record 46.5 inches (118 cm), 31 lb, 12 oz.
Snapper blues: Baby bluefish, 6 to 8 inches (15-20 cm).
Reproductive age: Maturity reached during the 2nd year of life.
Reproductive season: South Atlantic Bight, spring - March to May.
Middle Atlantic Bight, summer - June to September.
Texas, short spring and short fall spawn.
Eggs: Buoyant, spherical. 0.9 to 1.2 mm in diameter, pale
amber yolk, single darker amber oil globule.
Egg production: A 3 to 4 year old female produces
from 0.6 million to 1.4 million eggs.
Incubation: At 65 to 72°F (18.5 to 22°C), 44 to 46 days.

*P*recisely what triggers the bluefish's yearly migration is not known. Most of its kind spend the winter off the coast of South Florida and the Gulf Stream; they begin their northward trek in late January and early February. Temperature apparently plays a critical role in the fish's distribution, feeding, spawning and recruitment success (survival in sufficient numbers to replace the stock of year old fish). The initiation of migration, however, occurs at a time when the area's ocean temperatures fluctuate very little. Thus researchers believe that, in this case, temperature may have little influence in initiating the northward move. Instead, the season's change in photoperiod, increasing light intensity and lengthening days, seems to be the main

Juvenile bluefish over a sandy bottom, Smithtown Bay, NY. Long Island Sound.

stimulus.

Swimming in schools that have been recorded stretching up to 5 miles (8 km) in length, older bluefish travel offshore to near the junction of the continental shelf and the Gulf Stream. As they move from northern Florida to Cape Hatteras, spring-spawning females roll themselves to one side and extrude their eggs. Attentive males then release their milt and fertilization occurs by mixing. Released in spurts as they travel, approximately 0.6 million to 1.4 million eggs are spawned by a 3 to 4-year-old female.

North of Cape Hatteras, the spent adults gradually turn toward the shoreline and enter Long Island Sound, Narragansett Bay or make their way past Cape Cod and farther north.

> **Determining gender:**
> The sex of a bluefish cannot be determined externally. The male matures earlier than the female, but its eventual size is not an indicator of gender.

Younger, smaller bluefish tend to move into Albemarble Sound, Chesapeake Bay and Delaware Bay. Both groups remain in their chosen habitats until the end of the season.

A second coastal Atlantic spawning occurs from June to August. These summer-spawners are believed to also spend their winter off south Florida. They then migrate past Cape Hatteras to a point over the continental shelf, some 31 to 93 miles (50 to 150 km) offshore. Eggs and sperm are released between Cape Hatteras and Cape Cod, before the fishes turn westward toward land. Many enter Long Island Sound while others continue past Cape Cod.

Most species of fish, like bluefish, spawn in open water and abandon their brood to the mercy of predators, the elements and disease.

Incubation of the bluefish's eggs occurs as they drift in the currents. Two hours after fertilization, cell division becomes apparent. By 34 hours, the embryo's heartbeat is visible and about 8 hours later, the tiny creature can be seen thrashing around in its protected environment. Depending on the water temperature, the larval fish

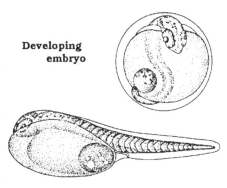

Developing embryo

Larval bluefish, 2.15 mm long
From Hardy, 1978.

emerges from its egg at 44 to 48 hours. The creature is barely .08 inch (2 mm) long at hatching; it carries a supply of yolk on its underside that is about half its length. Within four days, the larva has grown to 0.12 to 0.16 inch (3 to 4 mm) and its supply of yolk is exhausted. The tiny creature then begins to feed mainly on copepods. As it grows larger and makes its way toward the shoreline, however, it adds fish, amphipods and crab larvae to its diet. Spring-spawned juveniles arrive in the bays and estuaries of the middle Atlantic coast in late spring or early summer. Summer-spawned blues generally remain out at sea throughout the season; while offshore, they continue to feed mainly on copepods.

The spring-spawned young are about 1 to 2 inches (25 to 50 mm) long when they arrive along the middle Atlantic shoreline. They feed ravenously while inshore and by September, they attain a length of between 7 to 8 inches (175 to 200 mm). After having spent their winter off Florida, these 1-year-old fishes reach a length of about 10.2 inches (260 mm). The summer-spawned blues do not grow nearly as rapidly. They hatch later in the season, and, since they generally remain offshore, they cannot avail themselves to the abundance of food found in the estuaries. A 2-year-old bluefish weighs about two pounds, and by its 10th year it can reach 15 pounds. The record blue caught on a rod and reel weighed in at 31 pounds, 12 ounces.

The bluefish is a voracious predator that has been variously described as "an animated chopping machine" and as "the most ferocious and bloodthirsty fish in the sea." It is well known for slicing through a school of prey, leaving in its

Weight/size
1-pound - about 14 in.
2-pound - about 17 in.
3-pound - about 20-21 in.
4-pound - about 24 in.
8-pound - about 28-29 in.
10-12-pound - about 30 in.
From Bigelow, 1953.

wake a large number of partially eaten and maimed victims. The bluefish also has been known to attack seabirds that are resting on the water's surface and those diving for food. In turbid waters of Florida and North Carolina, marauding blues have been blamed for attacks on swimmers. A visual feeder, this streamline, muscular creature leaves its school and increases its speed as it bears down on its prey. When it is within inches of its victim, the fish drops its lower jaw and arches its head. Armed with sharp, conical teeth that line both jaws, the fish bites off a piece of its victim's flesh or swallows it whole. Once the chase is completed, the predator returns to the school.

Young bluefish feed on small shrimps, killifishes, silversides and a variety of other estuarine residents. As these predators grow in size, they seek progressively larger prey. Adults feed on a number of creatures including squid, crabs, butterfish, menhaden, shad, herring, hake and smaller members of their own species. Bluefish are predated by sharks, tuna, swordfish and humans. Only about 10 percent of bluefish are harvested by commercial fishermen; the others are taken by sports fishermen.■

CREATURES OF THE ROCKY INTERTIDAL AND ROCK REEFS

Rocks, pilings, shipwrecks, shellfish beds and worm reefs

3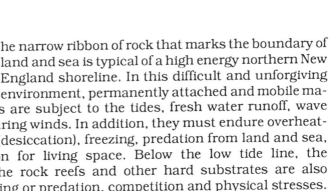

*T*he narrow ribbon of rock that marks the boundary of land and sea is typical of a high energy northern New England shoreline. In this difficult and unforgiving environment, permanently attached and mobile marine inhabitants are subject to the tides, fresh water runoff, wave action and scouring winds. In addition, they must endure overheating by the sun (desiccation), freezing, predation from land and sea, and competition for living space. Below the low tide line, the organisms of the rock reefs and other hard substrates are also exposed to grazing or predation, competition and physical stresses. On the other hand, the continually submerged structures are in a more stable environment. The rocks, wrecks, pilings and other hard structures provide their inhabitants with shelter from predators, a place to lie in ambush, a convenient site for feeding and growning and a breeding ground.

Physical and biological factors combine to separate the plants

79

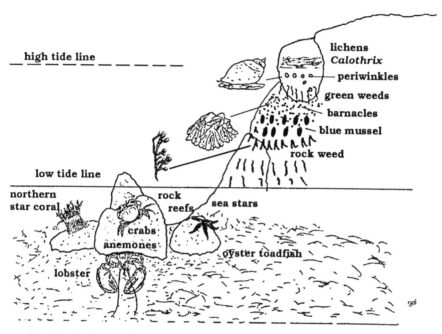

and animals living between the tides into zones where only certain species have managed to adapt. Such zonation is evident not only on the New England coast, but also at sites worldwide including the British Isles, South Africa and Australia. Along the Atlantic coast south of Cape Cod, MA, rock substrates gradually give way to mainly man-made structures and biologically created habitats. Man-made structures include jetties, groins, dock pilings, bridge abutments and shipwrecks. Biological structures include oyster beds and worm reefs. Though zonation is most obvious on the rocky New England shoreline, it is often evident (but usually less distinct) on pilings and other structures that rise above the surface waters. The populations living on these structures are similar to those of a rocky coast.

Plants and animals inhabiting the rocky intertidal zone increase in number and variety as one proceeds from the land to the low tide line. The residents of the upper reaches of the intertidal zone (littoral fringe or high intertidal) tend to be related to land-dwellers nearly as much as they are to marine inhabitants. This area is occasionally washed by the sea during spring tides and storm waves and by salt spray. It is populated by organisms such as the lichen *Verrucaria* and the cyanobacterium (blue-green algae) *Calothrix* which form the characteristic tarlike patches or bands seen high on the rocks. The

80

rough periwinkle, *Littorina saxatilis*, is one of the few mobile animals that penetrates this zone. The snail is easily able to withstand temperature extremes, and it apparently can survive up to a month without being moistened by salt water. It feeds on *Calothrix* and other microbial biomass.

Barnacles are generally the dominant species that occupy the area below the lichens and blue-green algae. Near the top of this zone, the barnacles are covered by sea water only during spring tides. Lower, however, the area is covered and uncovered with each tidal cycle. The barnacles then give way to blue mussels, and they, in turn, are replaced by rock weed (*Fucus* spp.), knotted wrack (*Ascophyllum nodosum*), and Irish moss (*Chondrus crispus*). Below the low tide line, the rocks are occupied by kelp (*Laminaria* spp.) and red weeds.

> **Predators of the New England rocky intertidal zone** (Menge, 1983)
> Predatory snails:
> Atlantic dogwinkle, *Nucella lapillus*
> Sea stars:
> Common sea star, *Asterias forbesi*
> Boreal Asterias, *Asterias vulgaris*
> Crabs:
> Green crab, *Carcinus maenas*
> Jonah crab, *Cancer borealis*
> Atlantic rock crab, *Cancer irroratus*

> **Zonation: Exposed structures - South Atlantic Bight** (Hay, 1988)
> Littoral fringe
> Blue-green algae
> Isopod (crustacean), *Lygia exotica*
> Highest point -barnacle zone
> Little gray barnacle, *Chthalmalus fragilis*
> Lower -barnacle zone: *Balanus* spp.
> Mussel zone
> Scorched mussel, *Brachidontes exustus*
> Oyster zone
> American oyster, *Crassostrea virginica*

> **Rocky habitats along the Gulf Coast:**
> The habitats are limited to biological and man-made structures. Organisms inhabiting the shoreline structures resemble those found on most rocky intertidals. Offshore, shipwrecks and oil drilling platforms serve as rock reefs habitats.

The numbers and distribution of intertidal residents can change significantly between a protected shoreline and one exposed to heavy wave action. Blue mussels dominate exposed northern New England shorelines, in the mid-intertidal zone. The intense battering by the waves apparently restricts predators as well as the growth of seaweeds. In more protected areas, predatory snails (Atlantic dogwinkle, *Nucella lapillus* [=*Thais lapillus*]), roam freely. They control the mussel population, allowing seaweeds to colonize the area. Oddly, this is generally not the case at very well protected sites. These tend to be dominated by barnacles and mussels with few seaweeds. Researchers believe that grazing by the common periwinkle, *Littorina littorea*, may play a key role in preventing the seaweeds from establishing themselves (see Common periwinkle, page 86).

Many of the invertebrates and fishes that inhabit the waters near the low tide line, forage in the intertidal zone during a rising tide.

As the tide ebbs, some animals remain behind in tide pools and others such as juvenile green crabs (*Carcinus maenas*) take shelter under rocks until the next tidal cycle. Each animal residing in the rocky intertidal or rock reefs has evolved its strategy for survival. Some, as will be seen, spend their adult life attached to or in association with hard substrates, while others remain there for just a brief portion of their life cycle. ■

The invertebrates

Blue mussel, *Mytilus edulis*

Habitat: On a solid substrate, intertidal and subtidal.

> Other common names: Mussel, sea mussel, common mussel, edible mussel.
> Phylum: Mollusca. Class: Bivalvia (clams).
> Order: Mytiloida. Family: Mytilidae.
> Geographic range: Arctic, Labrador to Cape Hatteras, North Pacific.
>   Europe: English Channel, North Sea,
>   Baltic and Mediterranean.
> Depth range: To 1,637 feet (499 m) in cooler waters.
> First spawn: Approximately at 1 year of age.
> Egg production: Approximately 15 million eggs per spawn.
> Number of sperm: 10,000 for each egg.

*T*iming is everything for a blue mussel spawn. Egg and sperm must be released simultaneously to assure fertilization. A few hours later, when the larvae hatch and begin their two to four week sojourn as plankton, food and conditions must be optimal to assure their growth and survival.

Preparation for spawning begins in the cold of winter when there is little available food or feeding activity. Thus, to produce its eggs or sperm, the mussel must draw on its reserves of glycogen (carbohydrate). With the approach of the spawning season in late spring, the mollusk continues to draw on its stores of glycogen and any food that it can filter out of the water. During this final stage, food availability and reserves, temperature, salinity, length of time that the mussel is exposed to air, and its hormonal cycle and genotype may all interact to ensure that the release of eggs and sperm is synchronized. The male is generally the first to spawn and the presence of sperm in the water apparently stimulates the female to release her eggs. Of the approximately 15 million eggs spawned by an individual female, a large percentage are never fertilized. In addition, an estimated 99 percent of the larvae are lost to predation or to the environment before they can metamorphose to a bottom existence.

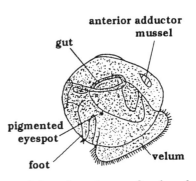

Internal anatomy of a larval blue mussel, veliger stage.
After Chanley and Andrews, 1971.

The larva begins its planktonic life as a non-feeding, spherelike creature whose surface is covered by hairlike cilia. As development progresses, the larva produces a thin shell and a ring of cilia (velum) surrounding its mouth region. Known as a veliger, this larval stage uses its cilia to gather food and propel itself through the water (also see Eastern mud snail, page 7). Though at the mercy of the currents, it can easily swim to the surface and when it retracts its velum, it sinks to the bottom. Older larvae move up and down in this fashion in direct response to changes in salinity.

With a decrease in salinity, characteristic of an ebbing tide, the older larvae drop to the bottom. As the tide rises, salinity increases and they swim toward the surface. The strategy is believed to help the creatures avoid being carried out to sea on an outgoing tide and allows them to penetrate farther up the estuary on a flood tide.

Near the end of the larva's planktonic stage, the creature develops a foot that allows it to crawl around on the bottom and search for the appropriate seaweed or a hydroid on which to settle. If it cannot find a suitable site, the tiny animal begins to swim once more. During this time of alternating between planktonic and bottom life, the larva is called pediveliger. Once settled, the larva loses its velum and secures itself to its chosen site by means of thin threads (byssus); it is then known as a plantigrade. At this stage the creature is about .010 inches (.25 mm) in length. It remains attached until it has grown to 0.4-0.6 inches (1 to 1.5 mm) and detaches itself. The tiny mussel is then swept along by the bottom currents until it finds an appropriate place to reattach.

The plantigrade generally chooses to settle among or on adult mussels or some other solid substrate. It attaches itself with byssal threads secreted from glands in its grooved foot. If the site proves unsatisfactory, however, the creature is capable of moving. Small mussels have been observed climbing the sides of aquariums. Using their foot, they attach a thread to the glass, release the old thread, and pull themselves up toward the new one. Repeating the process, the creatures can move up to the edge of the tank. In nature, a mussel moves itself by adjusting the length of its anchoring threads and/or by secreting new ones. In doing so, it can reposition itself to take advantage of prevailing currents or it can move away from accumulating sediments that threaten to bury it.

Blue mussel,
Mytilus edulis

The young blue mussel keeps itself meticulously clean. Using the licking motion of its tonguelike foot, it scrubs its shell. Gathered particles are then carried by the foot's two bands cilia to the mantle cavity. Once there, they are processed for consumption or rejection through the gills as are waterborne materials entering the creature's inhalant siphon.

The cleaning behavior of the young mussel also tends to prevent other mussels from attaching themselves to it. When an older mussel attempts to reposition itself, it often finds itself held in place by its neighbor's byssal threads. During a heavy set of new mussels (spat), the older

Anchoring byssal threads, as seen through a field microcsope, 15 X.

mussels can easily be smothered. But in the case of the young mussels, the spat are usually scrubbed off the shells. Thus, the spat must settle on the solid substrate between the mussels or on their byssus. The inability to find a suitable site on which to settle leads to huge losses among them.

Blue mussels often form large aggregations on wharf pilings and rocks that are exposed to heavy seas. It is believed that the strong waves help limit predation, but sheltered areas may also be host to large numbers of the bivalves. Most of these mussels, however, are anchored to each other and not to a solid substrate. Under those circumstances, storms may lift and tear away large portions of the beds. Predation may also reduce their numbers in such an area.

The loss of the blue mussel due to predation is greatest during larval stages. Once the creature has metamorphosed and grown a shell of 1.5 - 2 inches (4 to 5 cm) in thickness, only predators such as the blackfish, *Tautog onitis*, sea stars, predatory snails, American lobster, some species of crab and certain seabirds can successfully prey on it. Through its association with the frilled anemone, *Metridium senile*, however, the mussel gains some protection from its enemies.

The anemone is often found either attached to the bivalve's shells or on the solid substrate next it. If a common sea star, *Asterias forbesi*, comes into contact with the anemone as it explorers the mussel, it reacts immediately. Apparently stung by the anemone, the sea star withdraws. Left to its own devices, however, the blue mussel seems entirely defenseless. Yet, the blue mussel has evolved an unusual system of its own against the predatory snail known as the Atlantic dogwinkle (=New England dog whelk), *Nucella (=Thais) lapillus*. The dogwinkle detects its mussel prey using its well-developed sense of smell (chemoreception), but it must first explore its potential victim's shell before committing itself to attack. Using its accessory boring organ on the front of its foot and its filelike radula, the snail drills a tiny hole through the mussel's shell (see Atlantic oyster drill, page 90). It then gains

Frilled anemone (=clonal plumose anemone), *Metridium senile.*

Atlantic dogwinkle

Blue mussel bed, Darien, CT.

access to the flesh by inserting its proboscis through the borehole. Sensing the slow-moving snail as it explores the shell, the blue mussel is often able to ensnare the predator with its byssus. With the help of neighboring mussels, the dogwinkle is completely immobilized. The 20 or more threads that are attached to the snail's shell are retracted, and the unfortunate dogwinkle is flipped over and left to die. It has been estimated that about 30 percent of the dogwinkles within a mussel bed are similarly trapped.∎

Common periwinkle, *Littorina littorea*

Habitat: Intertidally on rocks, pilings, sand and mud.

Other common names: Periwinkle, winkle.
Phylum: Mollusca. Class: Gastropoda.
Order: Mesogastropoda. Family: Littorinidae.
Geographic range: Western Atlantic - Labrador to Maryland.
                      Europe - White Sea to Gibraltar.
Southward extension of its range:
                    Nova Scotia, 1840
                    Bay of Fundy, 1861
                    Provincetown, MA, 1870
                    New Haven, CT, 1879
                    Staten Island, NY, 1888
                    Cape May, NJ, 1928
                    Ocean City, MD, 1959
Size: South of Cape Cod 1 inch (27 mm),shells thinner.
      North of Cape Cod and in Europe, 1 1/3 inches (35 mm),
      shells thicker.
Unusual characteristic: Pearls have occasionally been found in
                      these snails on the coast of England.
Life span: South of Cape Cod 3-5 years.
           North of Cape Cod and in Europe, 10-20 years.

*T*o those who frequent the shores of the Gulf of St. Lawrence to New Jersey, the common periwinkle, *Littorina littorea*, hardly needs an introduction. Yet this, one of our most common snails, is only a relatively recent European immigrant.

    Littorina was first reported on the North America coast at Pictou, Nova Scotia, in 1840. By 1870, the snail had spread to the north shore of Cape Cod and nearly a decade later, it had found its way into Long Island Sound. In 1928, it was discovered on the intertidal rocks of Cape May, NJ, and by 1959, it had extended its southern range to Ocean City, MD.

    At first, it was speculated that the creature had probably crossed the Atlantic, accidentally or intentionally, aboard European vessels around 1840. During the early 1960s, however, discarded shells were found at Mimac Indian camp sites in Nova Scotia; radiocarbon testing placed their age around 1000 A.D.

86

Common periwinkle, *Littorina littorea*, Norwalk, CT.

The periwinkle is very common along the Norwegian shore-line. It is frequently found there in ancient refuse piles with the shell's of other edible mollusks. The Norse are known to have sailed from Greenland to Newfoundland and Nova Scotia around 1000 A.D., and the snail may have been introduced at that time. But its failure to spread from Nova Scotia until the mid-1800s, may have been due to regional ocean (surface) circulation of that time. Once conditions were favorable, the periwinkle began its southward invasion. Its arrival on the shores of New England was soon felt by the Eastern mud snail, *Ilyanassa (=Nassarius) obsoletus*.

An 1873 survey of the Cape Cod region described *Ilyanassa* as "...dominant on sand and mud flats, pilings, sea walls, salt marshes, and eel grass beds, and common on protected rocks, cobble beaches, and pilings." In 1899, naturalist F. N. Balch observed that the periwinkle was beginning to crowd out the Eastern mud snail at Cold Spring Harbor, NY. Two decades later, *Ilyanassa* had been displaced from 70 percent of its former habitat and then was confined primarily to the soft mud/sand flats.

During its reproductive season, the Eastern mud snail makes its way to solid substrate to deposit its eggs, and in some areas, it has continued to share its habitat with *Littorina*. When it encounters its competitor too frequently, however, *Ilyanassa* moves away. What precisely initiates the behavior is not yet understood.

Some have speculated that the Eastern mud snail's displace-ment might be in part a result of competition for food. The adults of both species consume a variety of small invertebrates and algae, and *Littorina* actively feeds on the mud snail's eggs. Both also feed on inorganic and organic matter in the sediments (detritus), but *Ilyanassa* works the sea bottom a great deal more as it scavenges for dead animals and plants. As it forages, the mud snail can decimate the local population of nematode worms and other small invertebrates. Its restriction to the soft sediments and its effects on others residing there may be, in reality, the greatest consequence of *Littorina's* invasion.

The periwinkle's spread was facilitated by its reproductive strategy. As spawning season approaches the male's reproductive appendage, located just behind his right tentacle, increases in length. During mating, the female is mounted from her right side and the male places his appendage underneath her shell, above her right tentacle. The process may take only about one or nine minutes. Often within just a matter of hours, the female releases her egg capsules in the receding tide. Resembling a World War I infantryman's helmet, each capsule contains from one to seven eggs (average three). The tiny structures are carried by the currents as the eggs develop into pear-shaped larvae (trochophore, see illustration page 66); the creatures then metamorphose and escape as free-swimming larval veligers (also see Eastern mud snail, page 7).

Aided by long shore currents and the Coastal Current which flows along the mid-Atlantic states, the periwinkle's larvae drift and continue their southerly invasion. Sea temperatures of 70°F (21°C) or more and the seaward swing of the Coastal Current just north of North Carolina, however, will probably help limit the snail's range.

After having spent a few weeks in

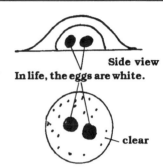

Side view
In life, the eggs are white.

clear

Common periwinkle's planktonic (drifting) egg capsule. Viewed from above. Drawn from author's photographs. The capsule measures 0.65 to 0.85 mm in diameter. From the side, the periwinkle's egg capsules resemble a WWI infantryman's helmet. Capsules contain one to seven eggs; the average is three.

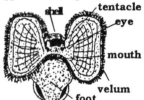

shell, tentacle, eye, mouth, velum, foot

Free-swimming veliger larva of the common periwinkle. The young veliger emerges from the flat side of the egg capsule some six days after the eggs are laid (Hayes, 1929).

Distinguishing the sexes and mating.
The male's penis is locate just behind his right tentacle; it is most obvious during the breeding season. When mating, the male places his reproductive appendage under the female's mantle, in front of her right tentacle. Mating pairs remain together for approximately one to nine minutes.

the coastal plankton, the larvae metamorphose and settle along the shoreline, below the low tide line. Then they begin their onshore migration. Using wave action and their own crawling abilities, the minute snails move into the intertidal zone. In time, the growing youngsters begin to occupy the upper part of the mid-intertidal zone while the larger, older snails generally reside at the edge of the low intertidal. If displaced above or below their chosen areas, the creatures invariably return to their home range. As if following the leader, individuals often use trails in the sand made by others of their kind.

The periwinkle's invasion also was helped by its remarkable capacity to adapt to the rigors of life

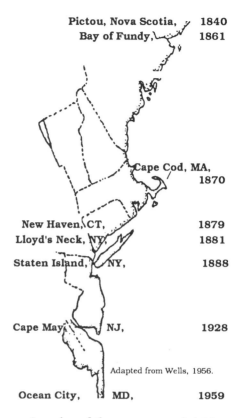

Invasion of the common periwinkle.

between the tides. The creature spends much of its time out of the water where it is subject to the changes of nature. During the summer, it has demonstrated its ability to withstand temperatures of as high as 111°F (44°C). As the heat rises to within several degrees of its lethal limit, the snail enters a reversible state of heat coma; it detaches its foot from the rocks or sand/mud bottom and extends it out from its shell. The periwinkle can tolerate winter temperatures of some 8.6°F (-13°C) by allowing ice crystals to form between its cells rather than within them.

The snail's use of almost anything as food has also helped it cope with its environment. In turn, the periwinkle serves as nourishment for another European invader, the green crab, as well as fish, whelks, gulls and ducks.

Two native species of *Littorina* share, in part, the same geographical area with the common periwinkle. These creatures, however, do not have a planktonic or a drifting larval stage, and thus they do not share the ability to easily expand their range. The rough periwinkle, *Littorina saxatilis*, which inhabits the upper intertidal and splash zones, caries its eggs in a brood sac; the offspring emerge as tiny, brown-shelled snails. The smooth periwinkle, *Littorina obtusata*, is commonly found on rockweeds, near the edge of the lower intertidal zone. It lays large egg masses on algae, and its young also emerge as fully developed miniatures of the parents. ■

Atlantic oyster drill, *Urosalpinx cinerea*

Habitat: Rock and shell bottoms, below the mid-tide line.

Other common names: Eastern oyster drill, oyster drill, American whelk tingle.
Phylum: Mollusca. Class: Gastropoda.
Order: Neogastropoda. Family: Muricidae.
Geographic range: Atlantic coast, Gulf of St. Lawrence to Florida.
  Introduced with oysters to the Pacific coast around 1870,
  and to the English coast around 1920.
Size: Maximum male length 1.14 inches (29 mm).
  Maximum female length 1.3 inches(33 mm).
  Newly hatched drills are purplish and about 1 mm in length.
Salinity limits: No less than 15 ppt; death reported at 12.5 ppt.
Crawling speed: Approximately 1 in/min (2.5-2.8 cm/min).
Rate of shell penetration: .03 to .05 mm/day.
Egg capsules: Vase-shaped, 15-.47 inches (4 -12 mm) high.
Egg production: Female produces 28-50 capsules per season.
  Average of 8.8 eggs/capsule.
Incubation period: Approximately 40 days.

*T*he oyster drill is an efficient predator; it can locate its quarry at a distance of several hundred feet! Extending its siphon, the snail turns into the current and swings its shell from side to side as it follows the scent of a prey. The drill usually attacks members of the same species within its immediate surroundings, but if the supply dwindles, it moves on to more plentiful prey. Its favorite foods consist of balanoid barnacles (see Rock barnacle, page 100) followed by newly settled oysters (oyster spat) and mussels. It also feeds on many other creatures including snails, bryozoans and small crabs.

Once it has located prey, the drill begins to explore the shelled creature with the front part of its foot (propodium). Tiny waves move across that portion of the foot that apparently allows the snail to feel the surface of its victim's shell. Occasionally, the drill stretches its

90

mouth open in what has been interpreted as a possible tasting behavior. Then, after some 10 to 30 or more minutes the snail begins its attack.

Atlantic oyster drill
From Galtsoff, 1964

As it starts to drill into its prey, the snail holds its foot firmly against the shell. The excavation involves the coordinated and repetitive actions of three of the snail's structures. Using its accessory boring organ (ABO), the predator deposits a strong acid on the surface of its prey's shell. The acid (pH 3.8-4) dissolves a small portion of the shell and the snail extends its file-like radula to scrape away and swallow pieces of the weakened structure. Housed in a tube-shaped proboscis, the radula is armed with rows of teeth that gradually wear away in the scraping process; new teeth continually replace the lost ones. The front part of the oyster drill's foot, the propodium, provides the creature with constant sensory information. It also eliminates seawater from the borehole that would otherwise dilute the acid. At the end of the drilling operation, the propodium breaks away any thin edges of shell and feeding begins.

The predatory snail inserts its proboscis through the borehole and tears away bits of the prey's flesh with its radula. In the laboratory, the snail has been observed feeding on an oyster for as long as two days. As the process continues, the victim's adductor muscle (the muscle that opens and closes the shell) weakens, and the shells part slightly. Attracted by fluids from the wounded animal, additional drills, crabs and fishes join the feast. Within as short time, the creature is consumed.

The rate at which the oyster drill penetrates the victim's shell depends on the predator's size, the hardness and thickness of the prey's shell, water temperature, salinity and other unknown factors. A drill apparently reaches the flesh of a mature oyster in about 2 1/2 days + - (0.3 to 0.5 mm per day). A rock barnacle (*Semibalanus balanoides*), which is penetrated through its moveable (opercular) plates, can be cleaned out in just 20 minutes! So it is little wonder that the snail keeps the barnacle at the top of its preferred food list. Nevertheless, the predator can be equally devastating to an oyster population. A study of the snail at Cape Cod, MA, found that in areas of dense snail infestation, oyster spat had little chance of surviving their first year. The thick shells of adult oysters help protect them from heavy losses due to snail attack; so does brackish water. The snail cannot withstand salinity much below 15 ppt, while the oyster flourishes in even less saline water. In its favored habitat, however, the drill reeks havoc on its victims from the time it crawls out of its egg capsule.

Following a winter of hibernation, the adult drill rises out of the sediments and climbs a rock or some other structure. Mating, which takes place as the temperature of the coastal waters increase, occurs at night. Sometime later, the female begins to lay capsules. Each filled with about eight eggs. Producing a total of about 28 to 50 capsules, the female deposits them in clusters, off the bottom, and close to a food source.

From Galtsoff, 1964

Atlantic oyster drill's egg capsule

The vase-shaped egg capsule is attached to rocks by its flat base. At the opposite end, an oval lid (operculum) seals the eggs in their protective environment. When the time of hatching approaches, an enzyme-like substance dissolves the operculum's seal and the creatures crawl out to the surface.

Over the next 40 days, the embryos develop within the confines of their protected environment. They then emerge as tiny replicas of the adult and immediately begin feeding on oyster spat, young hard-shelled clams or any other appropriate victim. Though the drills tend to be regionalized, they can, as juveniles, be carried to new grounds by crawling on floating debris that has settled to the bottom during a slack tide. Adults also have been observed riding on the backs of horseshoe crabs. ■

Oyster drill exploring the surface of an oyster.

Northern star coral, *Astrangia poculata*

Habitat: Rocks and other hard substrates. **(=A. danae)**

Other common names: Northern stony coral, star coral, cold water coral.
Phylum: Cnidaria. Class: Anthozoa. Order: Scleractinia. Family: Rhizangiidae.
Genus *Astrangia*: Approximately 30-36 living species, 20 fossil species.
Geographical range: Cape Cod to Florida. Gulf of Mexico.
Depth range: Low intertidal to 131 ft 40 m).
Temperature tolerance: 30 to 71.6ºF (-1.5 to 22ºC).
Salinity tolerance: 10 to 45 ppt for up to 36 hours.
Reproduction -sexes: Sexes separate, dioecious. All of the polyps in
                     a colony are generally of the same sex.
Egg production: Approximately 6,000 eggs per female polyp.
Reproductive season: Narragansett Bay, RI - August.
Zooxanthellae: Yellow-brown unicellular algae of the family *Dinophyceae*.

*D*ivers exploring Southern New England's rock reefs often encounter small white or brown colored, fuzzy looking "growths" on the sides of rocks. Though there may be other similar-looking creatures attached to rocks, the growths are frequently a species of stony coral.

The northern star coral, *Astrangia poculata*, is the only stony coral found in New England waters. It is a non-reef-building (ahermatypic) coral; the entire colony is often no larger than a silver dollar (see box below). The coral's ability to survive temperatures from just below freezing to approximately 71.6ºF (22ºC), has allowed it to establish itself in coastal waters from Cape Cod to Florida. Reef-building (hermatypic) corals require temperatures above 68ºF (20ºC). They are thus restricted to tropical and subtropical (e.g. Bermuda) waters. As will later be examined, however, there are other significant differences between *A. poculata* and reef-building corals.

A colony of northern star coral is composed of five to 30 individual polyps. At times, colonies of the polyps lie so closely together that they form large, slightly raised encrustations, several feet wide. Northern star colonies found along the east coast of Florida are generally branched and their polyps more commonly contain zooxanthellae than those from Narragansett Bay.

The colonial, lined anemone, *Fagesia lineata*: It is sometimes mistaken for northern coral by SCUBA divers. Block Island, RI.

93

Corals belong to the Phylum Cnidaria, a group of animals that also includes hydroids, jellyfish and anemones. The polyps, or living part of stony corals, closely resemble anemones in body form, but unlike anemones, they produce a calcium carbonate skeleton. Coral polyps and anemones have a ring of tentacles surrounding a mouth-like opening. The tentacles are armed with stinging cells (nematocysts) that fire barbed, harpoon-like projectiles and toxins into a prey. (The knobs on the ends of the northern star coral's tentacles house batteries of nematocysts that may be 29 microns [1 mm = 1,000 microns] in length.) Once a victim has been stunned or killed (usually small zooplankton), it is directed to the mouth by the tentacles and then swallowed.

The majority of New England's northern star corals are totally dependent on captured prey for their sustenance. In contrast, reef-building corals have an additional and more important source of nutrition. They have developed a mutually beneficial relationship with the algae, zooxanthellae. These symbiotic algae carry out their life functions within the tissues of their polyp host and get their energy from sunlight. The products of their photosynthesis (O_2, glycerol and fatty acids) are used by the polyps. In addition, the calcification of the corals' skeleton is accelerated considerably by its association with the symbiotic algae. Corals, in turn, provide a protected environment for the algae, and a ready source of CO_2, nitrogen and phosphorous (waste products). "On this symbiosis rest the entire biological productivity of the coral reef ecosystem" (Goreau, 1979).

The reliance of reef-building corals on zooxanthellae and the algae's need of sunlight for photosynthesis, restricts these corals to a maximum depth of about 328 feet (100 m). Most reef-dwelling corals harbor zooxanthellae, but far fewer of the non-reef-building corals have such a relationship. The northern star coral is a notable exception.

The northern star coral exists with or without the zooxanthellae. In Southern New England, the brownish colored symbiotic

Northern star coral with its polyps expanded, Fishers Island, NY.

94

and whitish non-symbiotic forms are often found side-by-side. Those harboring the algae grow faster and calcification of their skeleton is enhanced. The abundant animal plankton (zooplankton) at the northern part of the corals' range, however, allows both forms to thrive. Precisely why certain forms have the algae while others do not, is not

Northern star coral with its polyps partially retracted, off Asbury Park, NJ.

known. Furthermore, the eggs and larvae of symbiotic corals do not harbor zooxanthellae. Thus, "the symbiotic relationship must be re-established with each succeeding generation" (Froehlich, 1980).

In Narragansett Bay, RI, spawning occurs during August. As polyps prepare to release their eggs or sperm (sexes are separate), their bodies increase dramatically in size. Their tentacles, however, are not extended outward as in feeding; instead, they remain "short and stubby." Held tightly closed, the mouth (oral disk) becomes more prominent. Then, in quick contractions of the polyps' column, streams of eggs or sperm are released into the surrounding water. In a short time, the procedure is repeated several more times. Fertilization takes place by chance.

The corals' flat-shaped larvae (planulae) emerge from their eggs some 12 to 15 hours after fertilization. Propelled by coordinated beats of their hairlike cilia, the planulae mingle with the plankton population. Within what may be just a matter of hours, the creatures settle to the bottom and develop into tiny coral polyps. Asexual reproduction, known as budding, then begins.

Corals increase their numbers both sexually and asexually. As we already have seen during its sexual phase, the northern star coral produces eggs and sperm. Its asexual phase is marked by the formation of new polyps from a single (mother) polyp. As the newly settled polyp grows, it sends out bud-polyps that replicate themselves in the same way. Eventually a colony of some five to 30 or more polyps is formed, all the same sex. By the age of 1 to 3 years, the colony is already capable of sexual reproduction. ■

Common sea star, *Asterias forbesi*

Habitat: Shellfish beds, rock, gravel or sand bottom.

Other common names: Common starfish, Forbes' common sea
star, Forbes' Asterias.
Phylum: Echinodermata. Class: Stelleroidea.
Subclass: Asteroidea. Family: Asteriidae.
Geographic range: Gulf of Maine to Texas.
Depth range: Low tide line to 160 feet.
Salinity tolerance: 15-20 ppt.
Reproduction - sexes: Sexes separate - dioecious.
Planktonic larvae in Long Island Sound: July to September.

*E*xtending its tube feet in a series of coordinated movements, the sea star, *Asterias forbesi*, pulls itself across the ocean floor as it searches for its next meal. Its tube feet are arranged in four rows that extend from its mouth to the tip of its arms (rays). At the end of each tube foot is a sucker-like disc that allows the animal to climb any surface and cling tenaciously to its prey. The sea star moves in a random fashion, at an average rate of three to six inches per minute. In what might be described as a burst of speed for the creature, it can propel itself in a straight line at 1 foot in 52 seconds! Despite its slow progress and seemingly haphazard movement, the sea star can locate a northern quahog (hard-shelled clam) hidden in the sand/mud bottom. Apparently it does so with the aid of chemical cues.

Forewarned of its approaching enemy, also apparently by chemical cues, the quahog slows its pumping activity and decreases the rate of its oxygen consumption. But its efforts are sometimes for naught. Once directly over the clam, the sea star begins to dig it out with its arms; sand is carried away by the tube feet that function like a conveyor belt. As a pit is gradually formed, the sea star lowers itself and begins pulling the clam out onto the surface. It then wraps itself around its prey. Using its arms and tube feet, the predator gradually pries the shells apart. With an opening of about 1/250 inch (.1 mm), *Asterias* everts its stomach and squeezes it inside the clam. Aided by enzymes, it begins digesting its victim's tissues.

Northern quahog,
Mercenaria mercenaria.

96

The quahog sometimes evades its enemy by digging deeper into the sediments, but the process is painfully slow (23/64 in/min - 9 mm/min). The creature can easily lose in this deadly conflict. Other mollusks, however, have evolved what appears to be more efficient means of escape.

When approached by a sea star, the bay scallop quickly snaps its shells shut and ejects water in the process. In a series of such moves, the animal swims erratically through the water, bounces off the bottom and shoots off again.

The Atlantic surfclam has also learned to avoid the predator, but it makes its escape by jumping away from the sea star. A juvenile surfclam reacts quickly to the touch of the predator's tube feet. The creature retracts its siphons and ejects water from its shell. If contact with the predator is maintained, the clam extends its foot. Pushing against the bottom, it catapults itself through the water. (This behavior has not been observed in the adult.) Despite evasive measures, escape is not assured and the sea star can exact a heavy toll on all three species. The predator can also destroy large numbers of the most valuable shellfish, the oyster.

Atlantic surfclam, Spisula solidissama: Launching itself off the sea floor.

When oyster culture began in Long Island Sound during the mid-1800s, most of the activity was centered within coastal harbors, near the mouths of rivers. At these sites, low salinity minimized the sea star's intrusion. It was not until around 1882, when deep water grounds were stocked with oysters, that the full impact of the predator was felt. In 1887, the sea star destroyed an estimated $463,000 worth of shellfish, and the following year it was responsible for a loss of $631,500. The extent of the continuing damage prompted the Connecticut General Assembly in 1901, to enact legislation making it a crime to assist in the spread of the dreaded pest. It read: "...every person who shall willfully deposit or assist in depositing any starfish in any navigable waters of the State shall be fined not more than fifty dollars, or imprisoned not more than six months" (Coe, 1972).

Early oystermen attempted to rid themselves of "five fingers" as it was then called, by tearing it apart and throwing it back into the water. The missing parts were, however, only a temporary inconvenience for the animal. In time, it regenerated its arms and the lost parts to its central disc. By the late 1800s, mops came into use as a means of controlling the predator. Weighed down by chains or an iron bar, the mops were dragged over the shellfish beds and the entangled sea star was boiled and thrown back into the water. In the late 1930s, oystermen began using quicklime as another control measure. Today, a granular form of this substance is spread over the shellfish beds at a rate of one ton per acre. Though expensive, the procedure is effective and apparently does no harm to the oyster. Fishermen, however, have continued to rely heavily on the star mop to keep the creature in check.

Bipinnaria (early) larva of the common sea star. Drawn from the author's photographs.

The sea star is a prolific breeder. In Long Island Sound, its reproductive season begins when the water temperature reaches 59°F (15°C). In preparation for spawning, the males and females arch their bodies with only the tips of their arms touching the bottom. Waves of rhythmic contractions then sweep up from the tip of their arms to their bases, and eggs or sperm are expelled. Fertilization takes place in the water and the eggs develop into free-swimming larvae.

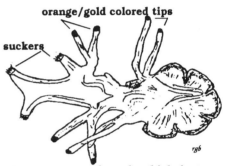

Common sea star larva, brachiolaria stage. The larva attaches itself to eelgrass, seaweeds, rock surfaces or any other suitable site using its suckers. It then metamorphoses into a tiny sea star. Drawn from the author's photographs.

The brachiolaria stage of the sea star's larva is one of the prettiest creatures that temporarily inhabits Long Island Sound's plankton. Trailing its gold-tipped arms, it is a delight to watch though a low-power microscope as it gracefully glides across a petri dish. The larva goes through several stages as it drifts in the Sound for three to four weeks. It then attaches it-

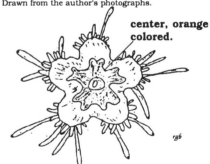

Newly metamorphosed common sea star. Drawn from the author's photographs.

98

self (set) to any available object and metamorphoses into a tiny sea star. Measuring about 1 mm in diameter, it immediately begins to feed on larval worms, young snails and small clams.

The rate of growth in the newly set sea star depends largely on its food supply. When provided ample nutrition, it can increase in size by as much 172 percent during its first four months of life. Under those circumstances, the animal reaches sexual maturity by the end of its first year. If food is scarce, the creature is known to prey on its own kind. *Asterias* can even live several months on a near-starvation diet, though in the process it generally shrinks in size. During the winter, the sea star moves a short distance offshore; the behavior may help it avoid being carried up on the beach in winter storms.

As an adult, the sea star has few enemies. A parasite has been found in males that may help control the creature's population, but there are scarcely any examples of marine animals that predate the sea star itself.

The American lobster has been observed feeding on the sea star, but the portly (nine-spine) spider crab may be one of the few creatures that regularly seeks it out. The crab's slender claws are easily inserted into a barnacle and by twisting these perfectly adapted appendages in opposite directions, it tears its barnacle-prey apart. The crab uses a similar technique when attacking a sea star.

The slow-moving spider crab has no difficulty in capturing its prey. It grasps a sea star's arms with both of its claws and, in a twisting motion, it tears a hole in the appendage; the injured arm is immediately cast-off (autotomized) and the victim makes its escape. Satisfied with its trophy, the crab then holds the arm like an ice cream cone and begins to snip tissues at the open end. ■

Redrawn after Aldrich, 1976.

Portly spider crab feeding on the arm of a sea star.

> **Why sea star and not starfish?**
> Other than sharing a common aquatic enviroment, sea stars are not at all related to fish. The name sea star is thus preferable to starfish.

99

Northern rock barnacle,
Semibalanus (=balanus) balanoides

Habitat: Intertidal rocks and subtidal in shallow water.

Other common names: Rock barnacle.
Phylum: Arthropoda. Subphylum: Crustacea. Class: Cerrepedia.
Suborder: Balanomorpha. Family: Balanidae.
Geographic range: Arctic to Cape Hatteras, NC.
Reproduction - sexes: Hermaphroditic, but cross fertilization is obligatory.
Reproductive season: Late autumn - October.
Egg production: Number produced varies from 400 to 8,000.
Larvae: Nauplii released in Long Island Sound January to March.
Life span: For most 3 years. Some may live to 5 or 6 years.

*T*he rock barnacle is a gregarious little crustacean that ordinarily inhabits rocks between the tides. Similar to other species of permanently attached (sessile) barnacles, the animal's habit of settling on the bottoms of oceangoing ships and small craft, have made it troublesome. In sufficient numbers, sessile barnacles affect vessels by slowing their speed, increasing fuel consumption and interfering with equipment. Some even promote metal corrosion. Their role as fouling organisms has been known for centuries; Aristotle in the fourth century B.C. observed that small fish (barnacles) could slow ships.

Barnacles are one of many representatives of the animal kingdom that make up the fouling community; organisms ranging from protozoans to chordates. Within an hour of being submerged, a surface begins to become colonized by bacteria. As the colony grows, a slimy and sticky matter (mainly polysaccharides) is secreted by the bacteria. This substance prepares the way for other fouling organisms.

On the East Coast, the hairlike, filamentous brown alga,

> **Copper-oxide antifouling paints:** These paints do not prevent the settlement of barnacle larvae. Their effectiveness is due to their toxicity which slowly kills the attached spat and newly metamorphosed barnacles.

Ectocarpus sp., is generally one of the first of its kind to grow on the slimy surface. It is followed by other species of algae and marine animals such as bryozoans, tunicates, sponges, tubeworms, mussels and barnacles. Though the slime layer is not necessary for the attachment of barnacles, the creatures show a strong preference for bacteria-prepared surfaces.

Settlement of the barnacles occurs during their final larval

100

(cyprid) stage. After hatching during late winter, the rock barnacle's triangle-shaped larvae (nauplius) drifts in the plankton, feeding on microscopic algae. Like other species of sessile barnacles, the creature passes through six nauplius stages and then metamorphoses into its non-feeding, cyprid stage. The clam-like larva propels itself with its six feathery legs (cirri), often with its first antennae (antennules) extended forward. As it searches for a place to settle, the creature responds to light, gravity and water currents.

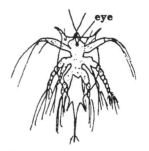

Barnacle's triangle-shaped nauplius larva.

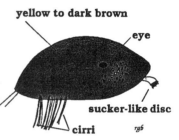

Cyprid larva of the northern rock barnacle. Drawn from the author's photographs.

The ends of the cyprid's antennules are equipped with sucker-like discs through which the animal secretes cement. It explores a possible homesite by crawling over the surface, alternately attaching and releasing its antennules. At times, the cyprid pivots on one antennule and surveys the surface with the other. Using chemical-tactile cues, the larva attempts to settle in a crack or pit, next to members of its own species. (A crack or pit gives the creature a better chance of surviving, and clustering, increasing the probability of successful reproduction.) If a site is not suitable, the animal can swim off to continue its search. When satisfied, however, the attached cyprid kicks its feathery legs to clean the area of silt. Then it pulls itself snugly up against the surface and tries to position itself toward incident light. Cement is spread around the antennules' discs. Depending upon the species, the cyprid metamorphoses into a young adult some four to 36 hours later.

In changing into the adult form, the rock barnacle produces a membranous base (basis) and surrounds itself with six separate and overlapping calcified plates. The top of the barnacle has two moveable calcified valves (operculum) that allow it to open and close. When feeding, the valves are parted and the creature extends its six feathery legs out of its shell. The appendages sweep the water for food.

The young rock barnacle grows rapidly during the spring of its first year. Its shell increases in size by continually adding calcified material along the edges of the plates and their interior surfaces. The membranous base grows in diameter as successive layers are laid, each a little wider than the last. As the animal grows within its shell, it sheds its outer skeleton (exoskeleton) like other crustaceans. By the fall of its second year, the rock barnacle is generally ready to reproduce.

Most stalked (eg. goose barnacle) and sessile barnacles are hermaphrodites, possessing functional male and female reproductive organs. Though there is evidence that a few species are capable of self-fertilization, mating with a neighbor is the normal occurrence; the northern rock barnacle always cross-fertilizes.

Feeding northern rock barnacle.

Rock barnacle mating occurs in October. In what might seem to be a Herculean effort, the creature extends its male appendage to more than twice its body length and deposits sperm within its neighbor's shell. The eggs incubate over the next three to five months and hatching is timed to coincide with a spring bloom of diatoms (phytoplankton).

By late March, Long Island Sound's intertidal rocks, shells and other hard surfaces are covered with newly set rock barnacle cyprids. Known as barnacle spat, large groupings of the creatures paint the structures in rusty-red hues. Many of the spat are lost to the scouring action of waves and shifting sediments. Some of the newly metamorphosed barnacles are smothered by encrusting sponges or bryozoans, or they simply loose out in the competition for space with blue mussels and rock weeds. When sufficiently crowded by others of their kind, the cone-shaped rock barnacles grow in long cylindrical shapes that are easily dislodged. Predators may also affect their population.

The Atlantic dogwinkle, *Nucella* (=*Thais*) *lapillus*, is the rock barnacle's greatest enemy. On a shoreline protected from heavy wave action, the snail can decimate the barnacle population. Its taste for prey, however, is not limited to the barnacle. It also seeks out blue mussel. The color of the snail's shell often reflects what it eats. Those that feed on barnacles tend to be grayish-white while mussel eaters are purple to brown (see Blue mussel, page 82). When attacking a barnacle, the dogwinkle stretches its foot over the top of the shell and secretes

Atlantic dogwinkle,
Nucella lapillus

102

a poisonous purple dye -purpurin- that kills the crustacean. The snail then inserts its proboscis and file-like radula between the relaxed valves and consumes its victim.

> **A barnacle is** ..."nothing more than a little shrimplike animal, standing on its head in a limestone house and kicking food into its mouth" Louis Agassiz, *circa* 1870.

Green crab, *Carcinus maenas*

Habitat: Rock, jetties and a variety of bottoms

Other common names: Shore crab, mud crab.
Phylum: Arthropoda. Subphylum: Crustacea. Class: Malacostraca.
Order: Decapoda. Family: Portunidae.
Geographic range: Mediterranean, English Channel, North Sea, Baltic, Iceland, Atlantic coast, Nova Scotia to Virginia, Australia. Introduced to the north and mid-Atlantic from Europe.
Depth range: Generally less than 20 feet. Occasionally to 650 feet.
Size(width): Male to 3.1 in (79 mm). Female to 3 in. (77 mm).
Sexual maturity: Male 1 in. (25-30 mm). Female 1/2 - 1 in. (15-31 mm).
Reproductive season: July to September.
Egg production: Females produce 185,000 to 200,000 eggs per clutch.
Life span: Females 3 years. Males 3 to 5 years.

*I*t is one of the most difficult times of its life. As is true for others of its kind, the green crab is trapped in the confines of its shell or external skeleton. It must free itself in order to continue to grow. Once the process of molting (ecdysis) or shedding has begun, however, the animal is virtually defenseless; it can barely move and it cannot feed itself. For any species of crustaceans held in an aquarium during this time, the stress of shedding is often the cause of death.

The first obvious sign of molting in the green crab occurs when its body begins to swell with sea water and a crack develops along its sides. Swelling increases and the crack trav-

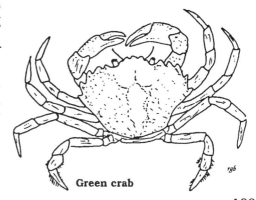

Green crab

103

The shell swings up between the two eyes.

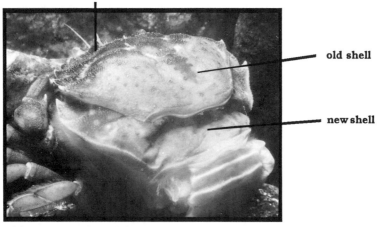

old shell

new shell

Molting green crab, Smithtown Bay, NY. Long Island Sound.

els along a suture line to the back of the shell and up to the eye sockets. Hinged on tissues between the eye sockets, the top of the old shell swings open and the crab slowly backs out trailing its flabby legs, claws and gills. At this point, the process has taken about one to three hours.

Once out of its old shell, the green crab continues to swell, increasing in size by as much as 30 percent. Then the shell starts to harden. Calcification or hardening of the new shell begins at the tips of the claws, mouthparts and walking legs. Depending on the size of the crab, water temperature, pH and calcium ion concentration, the process is completed within three to 16 days. If the crab loses a claw or leg during shedding, or from a confrontation, it can then regenerate the lost appendage. Shortly after the loss, a small bud appears on the stump. Growth of the appendage to its normal size, however, must wait for the next molt(s); the exact timing depends on the size of the crab, nature of the injury, food supply and water temperature.

The lifelong molting cycle also affects reproduction. Mating can only take place immediately after the female has shed, while she is still soft. A female green crab that is ready to mate sometimes displays her underside, and moves toward a male and touches him with one of her claws. If receptive, the male picks her up and carries her with her back touching his underside until she is ready to shed. Upon shedding a few days later, the male turns the female over and carries her with their undersides touching. The couple's abdomens are then hinged backward and for the next few hours, they mate.

Some time after mating, the female backs herself into a sandy bottom and lays 185,000 to 200,000 brightly colored orange eggs.

Glued to hairlike appendages on her abdomen, she carries them as they gradually change to a dirty-brown color. When it is time to release her young, the female pumps her abdomen back and forth and dispatches hundreds of mosquito-like larvae into the shoreline waters. As they drift in the currents, the larvae molt and grow into their next stages, and ultimately change into the crab-like megalopa stage (see megalopa, page 106). At the completion of this phase, the larvae sink to the bottom and change into tiny crabs.

Mating green crabs, Cape Ann, MA.

Female green crab with eggs, Sherwood Island State Park, CT. Long Island Sound.

The newly metamorphosed crabs take up residence along the shoreline where they remain until spring. In Long Island Sound, the juveniles and small Say mud crabs are generally the only species of crabs found intertidally during the winter. Hidden under rocks near the low tide line, the young green crabs are camouflaged in patterns of black, white, green and/or red; they blend in well with their background. Pigment cells (chromatophores) and/or a pigmented layer in their shell, allow the youngsters to change color in response to their background, temperature and time of day (solar-day rhythm). Their shell becomes lighter in color at night or with an increase in temperature; in daylight, the shell darkens. Adults are also capable of changing color, but the process that requires as little as a half-hour in juveniles, can take up to several days in the older crabs.

During the spring, the adults return from their offshore wintering grounds. Some make their home in the intertidal zone and emerge to feed as the rising waters wash over their refuge. Others remain below the low tide line and move shoreward to forage during a high tide. For both groups, however, the period of greatest feeding activity occurs at night, on a high tide.

The green crab has a varied diet. It consumes some algae but it tends to feed mainly on snails, bivalves (clams), worms and a variety of small crustaceans. It locates its prey by chemical cue (chemoreception).

Crab-like megalopa stage

When it "smells" its prey, the hungry crustacean's first antennae (located closest to the animal's midline, between the eyes) begin to flicker rapidly. Then, as if licking its chops, the crab's mouthparts (third maxillipeds) move together and back and forth. Following chemical cues, it begins its search in a seemingly erratic pattern. As it moves across the sea floor the crustacean sometimes travels past its prey. "Smells" produced by its victim, however, redirect it, and the crab finally locates its meal.

The crab easily devours a small periwinkle. Using one of its claws, it crushes and consumes the snail in about three minutes. But a medium-size periwinkle offers more of a challenge. The crustacean is generally unable to crush the snail's shell, and so begins to chip away at the opening. Once past the periwinkle's lidlike operculum, the crab is able to grasps its prey's body with one claw and tug from the opposite direction with its other claw. When attacking an Eastern mud snail, the green crab chips away at the pointed end (apex) of its victim's shell. In the process, it scatters bits of its meal over the bottom, triggering an escape response in other mud snails (see Common periwinkle, page 86, and Eastern mud snail, page 7). Though the green crab readily preys on periwinkles and mud snails, it apparently favors clams.

During the early 1870s, the green crab ranged from Great Egg Harbor, NJ, to the tip of Cape Cod, MA. By 1905, it was reported as far north as Casco Bay, ME. Inadvertently carried aboard lobster and sardine vessels, the crustacean eventually extended its range into Nova Scotia. The gradual warming of the coastal waters, however, ultimately allowed it to succeed in permanently establishing itself. Following the spread of the green crab north of Cape Cod, landings of soft-shell clams decreased dramatically in that same area. Much of the decline is suspected to have been due to predation by the crustacean.

Portly spider crab, *Libinia emarginata*

Habitat: Rock, sand, mud.

Portly spider crab, *Libinia emarginata*
Other common names: Common spider crab, nine-spine spider crab.
Phylum: Arthropoda. Subphylum: Crustacea. Class: Malacostraca.
Order Decapoda. Family: Majidae.
Geographic range: Nova Scotia to the western Gulf of Mexico.
Depth range: Shoreline to 131 feet (40 m). Occasionally to 407 ft (124 m).
Size: Length - male 4.9 in. (124 mm); egg-bearing female 2.6 in. (69 mm).

Closely related species: Longnose spider crab, *Libinia dubia*.
Other common names: Six-spine spider crab.
Class: Malacostraca. Order Decapoda. Family: Majidae.
Geographic range: Cape Cod, MA, to southern Texas. Bahamas and Cuba.
Habitat: All bottom types.
Size: Length, male 4 inches (102 mm); egg-bearing female 2.9 in. (74 mm).

*T*he young spider crab moves sideways in an awkward gait as it approaches a large stand of seaweeds. Having selected Irish moss (*Chondrus crispus*), the creature snips off a piece with its long, slender claw and carries it to its mouth. Then, using its mouthparts, the crustacean roughens the seaweed's edges in preparation for attaching it to its shell.

The crab's rostrum - the spine that extends forward between its eyes - and other parts of its external skeleton have tiny hook-like hairs (setae) that hold the seaweed in place. Once the material is ready, the creature rubs it across the back of its shell in an attempt to entangle or impale it on the setae; the process is part of the juvenile's decorating or camouflaging behavior.

To make itself less obvious, the crustacean often uses a variety of other materials including sponges, bryozoans, ascidians and hydroids. In the right setting, a well-decorated spider crab is difficult to spot. It sits on the bottom with its yellowish-colored claws tucked under its shell and often the only sign of life is the occasional flicker of the animal's antennae. As it moves from one place to another, however, its dress does not necessarily match the surroundings. A crab that is partially decorated with red beard sponge (*Microciona prolifera*) is quite conspicuous in a bed of green-colored algae. Nevertheless, the creature does not attempt to readjust its camouflage. The fact that it simply does not look like a crab is thought to sufficiently shield it from some of its enemies.

107

Young portly spider crab. Note the short rostrum (the spine beteen its eyes). Decorations on both crabs were removed to make photo identification easier.

Young longnose spider crab. Note the deeply forked rostrum.

The portly spider crab, *L. emarginata* and the longnose spider crab, *L. dubia*: The young of these two species are difficult to tell apart. The rostrum of the young *L. dubia* is longer and deeply forked in comparison to that of *L. emarginata*. Along the middle of its back (dorsum), *L. dubia* has six spines; *L. emarginata* has nine.

108

It is only the young crab that decorates itself. Judging from the combative spirit of the adult, it apparently does not need camouflage. A SCUBA diver approaching an adult buried in the sand is often challenged by the creature. Emerging from the bottom, the crab charges toward or retreats from the intruder with claws outstretched. Any attempt to pick it up is usually rewarded with a

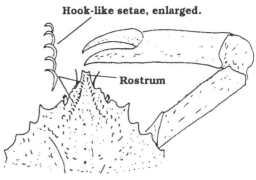

Hook-like setae, enlarged.

Rostrum

Portly spider crab.
Hook-like hairs (setae) covering the spider crab's rostrum.

sharp pinch from the animal's slender, but strong claws.

The portly spider crab, *Libinia emarginata,* and its close relative the longnose spider crab, *L. dubia,* share similar habitats and geographical range. The portly spider crab is the most common large spider crab on the shrimping grounds of the western Gulf of Mexico. The longnose spider crab is the most common spider crab in Tampa Bay, FL. Juveniles of both species decorate themselves and have developed a remarkable relationship with certain species of jellyfishes (scyphozoans). They ride inside the bell of these stinging creatures with no apparent harm to themselves. Young longnose spider crabs are often found in pits, excavated in the oral arms (scapulets) of the cannonball jellyfishes (*Stonolophus meleangris*). The crabs not only get free transportation, they also feed on its host's tissues and tentacles.

One of the adult portly spider crab's dietary habits is no less surprising: it feeds on the arms of the sea star, *Asterias forbesi* (for a description, see Common sea star, page 99). Like the juvenile, the adult also has evolved an unusual behavior of its own, known as podding.

During a late 1980s bottom survey of western Long Island Sound, NOAA vessels noted large, unexplained mounds in areas of otherwise uninteresting topography. The mounds later disappeared. A few years earlier, University of Connecticut scientists had discovered large aggregations - pods - of molting portly spider crabs in Fishers Island Sound, NY. The podding crabs were found on featureless, muddy bottoms, at depths of 9.8 - 16.4 feet (3 to 5 m), and 75.5 - 82 feet (23 to 25 m). The mysterious mounds of Long Island Sound also proved to be molting spider crabs.

In what may be a yearly autumn event, the portly spider crabs come together in large groups. They form pods of two to five animals deep that can cover a circular area of some 200 ft² (62 m²). Most individuals found on or near the pods have recently shed their external skeleton and are still soft. Apparently, the gathering is not related to reproduction; the female mates while her shell is hard - during an intermolt period. Thus, the delay of molting until the autumn cold and the formation of large groups is probably a protective strategy; by this time in the more northern waters, most predators have already migrated offshore for the winter. After podding, the spider crabs disperse. During the winter, they are often found buried in the sand with only the surface of their shell left exposed.

In preparation for reproduction, the male holds the female with her underside in contact with his own. Grasping her abdomen with his claw, he bends it backward and lowers his abdomen. After mating has taken place, the male abandons the female. Later she extrudes her eggs.

At Woods Hole, MA, egg-bearing portly spider crabs have been recorded from May to September. On the shoreline of North Carolina, they have been observed from June to August. When first extruded, the eggs are bright red/orange, but with further development and absorption of the yolk, their color turns to brown.

As the female's time to free her larvae nears, a male sometimes approaches her to assists. With their shells touching, the male reaches for her abdomen with one of his fifth walking legs and inserts the tips under the flap. Holding the appendages in that position, the pair retreats to some sort of shelter and the male fends off others of his sex with his claws outstretched. Within a short time, the female releases her larvae and the pair disengages.

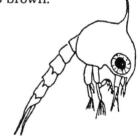

Larval longnose spider crab, first zoea.

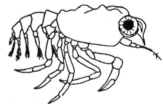

Larval longnose spider crab, megalopa.

Redrawn after Sandifer, 1971.

In an aquarium, females have been observed extruding a second brood some 12 hours later. The hatched larvae join the plankton where they pass through mosquito-resembling zoeal stages and a crab-like megalopa stage before metamorphosing into bottom-dwelling crabs. ■

American lobster, *Homarus americanus*

Habitat: Rock reefs, soft, compact mud, Pleistocene clays.

Other common names: Maine lobster, northern lobster.
Phylum: Arthropoda. Subphylum: Crustacea. Class: Malacostraca.
Order: Decapoda. Family: Nephropidae.
Geographical range: Labrador to Cape Hatteras, North Carolina.
                 Most common from Nova Scotia to New York.
Temperature tolerance: 30 to 82.4°F (-1 to 28°C).
Salinity tolerance: Down to 8 ppt.
Normal color: Greenish brown.
Genetically blue lobster: 1:30 million.
Size: Male record, 42 pounds (19.25 kg), 25 in (63.4 cm) long.
    A 15 pound male is approximately 25 years old.
    Female record, 18.4 pounds (8.35 kg).
Size at maturity: Males, 1.6-1.8 in (40-45 mm) CL.
               Females in Long Island Sound (LIS), 2.8-3 in (70-74 mm)CL.
               Females in Maine, 3.2 in (90 mm) CL.
Egg production: 3,000 to 115,000 eggs per clutch.
Larval lobsters: Peak hatching intensity in June and early July in LIS to
           early August in Maine.

*F*rom Labrador to North Carolina, the Atlantic is host to one of the more delectable creatures harvested from the sea, the American lobster. In some areas, small juveniles can be found under intertidal rocks, especially during a very low spring tide. Larger juveniles and adults inhabit the waters from below the low tide line to depths of some 1575 feet (480 m). Just where they take up residence depends a great deal on water temperature and the nature of the sea floor.

Near the shoreline, lobsters often dig sand from under flattened rocks and boulders, or they occupy available burrows or crevices. They tend to prefer a refuge sheltered from light. The creatures also tunnel below blue mussel beds, stands of eelgrass, or man-made structures such as shipwrecks and artificial reefs. In many harbors and estuaries from Nova Scotia to western Long Island Sound, soft, compact muds also serve as burrow sites. Young lobsters residing in Long Island Sound, also excavate shelters under large clumps of boring sponge (*Cliona celata*).

The creatures shape their home so as to maintain physical contact with the structure's walls and roof. Excavation is accomplished in a number of ways. Sediment is dug with the walking legs and scooped up with the mouthparts, tail fan or legs. The claws serve as rakes and bulldozer blades. Loose materials are swept out of the shelter using the fanning action of the swimmerets. Many of the burrows are produced with a main entrance and a smaller rear exit.

111

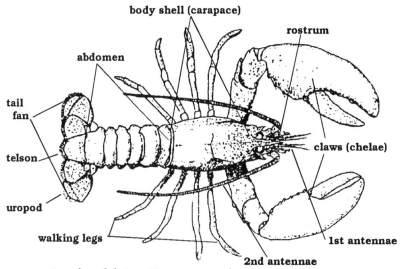

body shell (carapace)

rostrum

abdomen

tail
fan

telson

uropod

walking legs

claws (chelae)

1st antennae

2nd antennae

American lobster, *Homarus americanus*
Modified from Long Island Sound: An Atlas of Natural Resources, 1989.

The opening at the rear allows for aeration and hasty departures.

Offshore lobsters live on the outer edge and upper slopes of the continental shelf. They frequently excavate their shelters in 10,000-year old Pleistocene clay. Dug in close proximity, these Pueblo Village-resembling burrows are home to a variety of other creatures including shrimps, crabs and tilefish. Young lobsters also take refuge in shallow craters next to mud anemones (*Cerianthus borealis*) while larger creatures additionally produce bowl-shaped depressions in clay-mud sediments. The scarcity of suitable shelters, however, often forces offshore lobsters to share their residence with others of their kind.

Coastal lobsters are usually solitary. They do cohabitate for mating and, in the cold of winter, some burrows are occupied by more than one animal. During this season, the creatures often take refuge in a mud burrow, blocking its entrance with mud and debris. They are said to remain inside for several weeks or months. In deeper water, coastal lobsters continue to move about, but they are sluggish and apparently do not feed very actively.

Throughout the warmer months, most inshore lobsters remain in their shelter during the day. They emerge after dusk to search for food. Juveniles with a body shell length (carapace length =CL)[1] of

---

[1] The body shell of the lobster is called carapace. The legal length (carapace length) of a lobster is determined by measuring the distance from the rear margin of the eye socket to the rear margin of the carapace.

Pueblo Village-burrows excavated in 10,000-year old Pleistocene clay.
Redrawn by Skip Crane, after Cooper, 1980.

less than 1.4 inches (35 mm) rarely leave their refuge. Once they have reached a length of about 1 1/2 inches (40 mm) CL, however, most already have started exploring the immediate vicinity of their burrow. Nocturnal foraging begins when the animals have grown to about 1.8 inches (45 mm) CL.

On an overcast day or at night, males can often be seen in their burrow with their antennae and claws extending beyond the entrance. A narrow beam of light shined in front of the creatures can entice them to move farther out to investigate. If pointed directly into the eyes, however, the creatures reel backward deep into the den. When mature males leave their burrow, they generally do so to hunt for food and spend time exploring their surroundings. Females usually leave only to feed. As they forage, the majority of these creatures apparently remain within 50 to 100 feet of their burrow.

Lobsters encountering a diver while outside of their residence, frequently walk directly away from the intruder. At other times, they raise their claws outward (meral spread) and walk backward (see photograph, page 118). If sufficiently alarmed, the crustaceans flip their tail and swim backward for a distance and they occupy any available burrow. After their nightly outings, the crustaceans usually return to their own residence or one nearby. Some, however, are migrants that travel in a specific direction over a distance of one half to a mile each night. Others move mainly in response to environmental changes.

> **Lobster habitats in Long Island Sound (LIS):**
> Eastern LIS: Major habitats consist of boulders and crevices under rocks.
> Western LIS: Major habitats consist of soft cohesive muds.

During heavy wave action, inshore lobsters often relocate to deeper water. The summer warming at the shoreline prompts western Long Island Sound's lobsters to seek deeper, cooler waters. They return to the river estuarine areas in the fall. Approximately 5 percent of Eastern Long Island Sound's population spends the warmer months 80 or more miles out on the continental shelf. Offshore lobsters also migrate.

Feeding:
Juvenile and adult lobsters scavenge, but they are mainly predators, catching live prey. They feed on crabs, sea urchins, mussels, periwinkles, polychaete worms, sea stars, fishes and sea weeds. They avoid any rotting animal matter. The creatures sometimes hoard their food, especially when they have gone without for a period of time.

The change of season heralds the migration of about 20 percent (some estimate the migration at 30 to 40 percent) of offshore lobsters to the warming coastal waters. With their claws held forward, senses alert and tail extended, the lobsters follow an unseen trail. The animals propel themselves forward with strong fanning movements of their swimmerets and their walking legs. Rock reefs, crevices and contours must sometimes be circumvented but the animals immediately return to their original path. A few individuals are known to move at speeds of up to 6.3 miles (5.5 nautical miles, 10.14 km) per day[2], but most probably cover only about 1 to 1.5 miles per day. The creatures have been tracked travelling distances of up to 214 miles in 71 days. It is believed that they make the trek to molt and reproduce. They return to their offshore habitat in the early fall.

Molting or shedding is an integral part of the lobster's life cycle. Its skeleton is external and like other crustaceans it must molt to grow (see Green crab, page 103). During its first year of life, the creature sheds as often as ten times. In its second and third year, it may molt three or four times and twice in its fourth year. It then generally molts only once a year.

Temperature regulates the frequency of molting and thus growth. Other factors affecting the process, however, include nutrition, loss of a claw or another appendage, habitat and social interaction. Low temperatures increase the length of time between molts (intermolt period). When the water temperature drops to 41°F (5°C), molting stops. Higher temperatures decrease the time between molts. When lobsters are raised at a constant temperature of 72°F (22°C), they can reach legal size within two years.

Inshore lobsters normally attain sexual maturity by their sixth year (fifth year in Long Island Sound). In the cold waters of the Gulf of Maine, they may take seven or more years to reach that stage. Those living at the head of canyons and the outer edge of the

---

[2] 1 nautical mile = 6,080 feet or 1.15 statute miles.

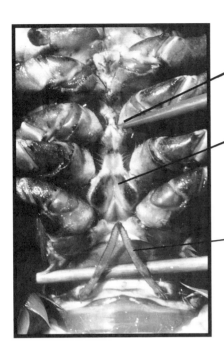

Female ventral (underside) view:
The female extrudes her eggs from tiny holes at the base of her second pair of walking legs.

Seminal receptacle:
The seminal receptacle is the shield-like, bright blue structure that lies between the last two pairs of walking legs.

First swimmerets:
The female's first pair of swimmerets (pleopods) are slender and soft.
The abdomen of a female about to extrude her eggs is dark.

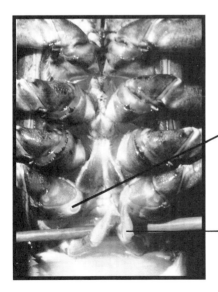

Male ventral (underside) view:
The male's claws are larger than a females.

Holes at base of last pair of walking legs: Long, gelatinous spermatophores (containing the sperm) are extruded from the tiny holes in the base of the male's last pair of walking legs.

First swimmerets:
The male's first pair of swimmerets (pleopods) are slender, rigid and grooved.

115

continental shelf are able to roam and feed throughout the year. They mature in three to four years.

The majority of females mate within 24 hours of molting, a time during which their shell is still relatively soft. Researchers have found, however, that they can also mate between molting periods, when their shell is hard. As soon as they carry sperm, the females are usually no longer receptive. Males can recognize these individuals and distinguish between mature and immature females. They seldom attempt to mate with non-receptive females.

Males may mate at almost any time of the year, even in very cold water. In laboratory experiments, a single male was observed mating with nine females in just 11 days. But individually, virility is quite variable. Some may not be capable of reproduction during the same year in which they molt. Size also makes a difference. Smaller mature males generally cannot mate with females that are much larger than themselves.

A female senses her approaching molt. Approximately seven days before its onset, she leaves the solitude of her burrow and begins searching for a mate. Researchers believe that she may use chemical cues (sexual pheromones) in her quest. If a male is suitable, usually a large dominant male, she approaches his den. At first, he may respond by raising his claws in a threatning gesture and attempt to push her away. She generally answers the aggression by lowering her claws and pushing them against him. The boxing match may go on for hours. Eventually, however, he relents and allows her to enter his den. Over the next few days, she regularly leaves the den only to return sometime later. Each time she approaches, the two greet each other by touching antennae and boxing. As the time of her molt draws closer, however, she spends longer intervals inside the burrow. About a half hour before the shed's onset, she often raises her claws and touches her mate's head, in a behavior is called knighting. Molting then begins with the male standing guard.

Her body swells as water is absorbed in the tissues. The defenseless female then rolls to her side and the back (dorsum) of her body shell swings up. The head's appendages are first to be with-

Female extruding, fertilizing and attaching her eggs.

Drawing by Rebecca Kelm

Female releasing her brood.

Drawing by John Kelm

116

Larval lobster

Stage I
0.3 in (7.5-8.03 mm).
Stage 1-5 days in length.
Note absence of swimmerets.

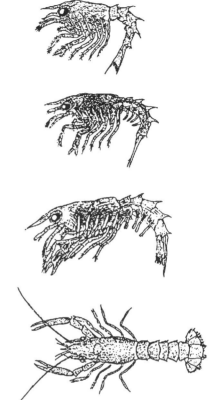

Stage II
0.3 -0.4 in (8.3-10.2 mm).
Stage approximately 2 to 7 days.
Note swimmerets and develop-
ing large claws.

Stage III
0.4-.047 in (10-12 mm).
Stage some 5 days long.
Tail fan completed, claws
larger and swimmerets have
short hair-like setae.

Stage IV
0.43-0.55 in (11-14 mm).
Stage last some 10-19 days.
Resembles a miniature
lobster and takes up a
bottom existence.

From Herrick, 1911.

drawn. These are followed by the appendages in the back part of the body shell (thorax) and finally the abdomen; the process takes 12 to 15 minutes. Mating takes place some 30 minutes later.

The male carefully rolls his mate over on her back, and the female stretches her claws out forward and extends her abdomen. Standing over her, head to head, the male picks her up and directs his sperm into her receptacle. Copulation is completed in less than a minute.

The sperm is stored by the female for about one year before she fertilizes her eggs; sperm can apparently remain viable for up to three years. When she is ready to extrude her eggs, the female rolls herself over on her back. From that position, she lifts herself up on her claws, and holds her body shell up at an angle of about 40⁰. She then curls her tail upwards to form a cup. Using one of her last pair of walking legs, she breaks the sperm receptacle's seal and begins to extrude the eggs. The stream of eggs flows from holes in the base of

117

Young American lobster, Newport, RI.

her second pair of walking legs. Aided by gravity and the beating of her first pair of swimmerets, the eggs tumble downward across the sperm receptacle where they are fertilized. As they reach the abdomen, they attach to the swimmerets.

The female carries her brood for about a year before they are ready to hatch. When first extruded, the eggs are dark green to black but as they near hatching they turn a golden brown with the eyespots of the developing larvae appearing blue. Possibly to protect her larvae from visual predators, she releases them at night. Turning herself into the current, the female stands on the tip of her walking legs and arches her abdomen upward. She then fans her swimmerets rapidly. The action sets her hatched larvae free, and they rise toward the surface where they join other zooplankton. Over the next few days to three weeks, she continues the process until all of her larvae are released. Precisely how long this takes depends on the water temperature.

The larvae pass through three mosquitoe-resembling stages and a fourth lobster-like stage before completing their lives as zooplankton. Very few of them survive to begin their lives as bottom-dwellers. In Long Island Sound, low visibility helps the lobster larvae avoid visual predators, but hunters are not their only hurdle. To remain alive, they must endure a long list of obstacles that includes extremes of temperature and salinity, disease and comb jellies (ctenophores). As adults, they face predation from sharks, Atlantic cod, wolffish, goosefish and people's insatiable appetite for their tender flesh. 🦞

Lentil sea spider, *Anoplodactylus lentus*

Habitat: Found among the hydroid *Eudendrium* spp., *Bugula turita* and other fouling organisms.

Other common names: Sea spider.
Phylum: Arthropoda. Class: Pcynogonida.
Family: Phoxichilidiidae.
Geographic range: Bay of Fundy to Caribbean. Lower
    intertidal to deep water.
Size: Body length to 1/4 inch (6 mm).
    Leg span to 1 1/2 inch (38 mm).
Reproduction - sexes: Sexes separate -dioecious.
Reproductive season: Mid- to late-August in the
    northern part of its range.
Life span: Unknown.

*A*wkwardly raising its long, spindly legs high above small obstructions, the lentil sea spider resembles an outer-space creature of Hollywood fame. The animal is purple, but its diminutive size and slow movements make it difficult to spot even against a light background. It belongs to a group of animals known as pycnogonids, or sea spiders. Worldwide, there are some 1000 species of sea spiders in 84 genera. The genus *Anoplodactylus*, to which the lentil sea spider belongs, has 110 known species.

The pycnogonids get their common name from their spider-like appearance. Some investigators believe them to have arisen from a line of marine arachnids (spiders, scorpions, ticks, mites, etc.) that never left their aquatic environment. Others, however, are skeptical of any such connection and the sea spider's ancestry remains uncertain.

The creatures inhabit the low tide line to the abyss, from the

Lentil sea spider: It is often found among hydroids or the bryozoan *Bugula turrita* in Long Island Sound.

polar regions to the tropics. The lentil sea spider's geographical range extends from the Bay of Fundy to the Caribbean. It makes its home from the low tide line to a maximum depth of about 900 feet (275 m). With a leg span of up to 1 1/2 inches (38 mm), it is regarded as a giant among those of its kind residing along the Atlantic shoreline; the majority are

119

microscopic in size. In comparison to the lentil sea spider, however, abyssal species are gargantuan. They are known to grow to a leg span approaching 29.5 inches (75 cm)!

Some sea spiders have five or six pairs of walking legs, though most, including the lentil sea spider, have four. Each of the lentil's legs has nine segments, the last of which is a terminal claw. Resembling a small hook or sharply pointed toenail, the structure allows the animal to cling to nearly any surface. The male has an additional pair of appendages known as ovigerous legs, which it uses to carry its eggs.

In the northern part of its range, the lentil sea spider's reproductive season begins mid to late-August. The male approaches his mate from behind and climbs up her back. Moving forward, he clambers over the top of her head and positions himself facing backwards on her underside. Then, as the female releases her white eggs, he rolls them up it a cohesive mass with his ovigerous legs. The eggs are apparently fertilized as they are extruded. Once the process is complete, the male uses the same appendages to hold the brood during incubation. In Long Island Sound, male lentil sea spiders have been observed carrying eggs from August to mid-October.

Sea spiders typically hatch as

Ringed sea spider
Tanystylum orbiculare.
Length: 1/16 in., 2 mm.
Found on pilings associated with hydroids, ascidians.
Range: bay of Fundy to Brazil.
After McCloskey, 1973.

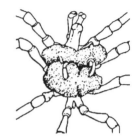

Male lentil sea spider carrying an egg mass.
From Cole, 1901

Protonymphon larvae of the sea spider *Paranymphon spinosum.*

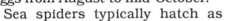

protonymphon larvae. These strange-looking creatures have three pairs of appendages. Two are walking legs and the other pairs are armed with claws. Most (if not all) larval pycnogonids have no planktonic or drifting stage, allowing them little opportunity to extend their range. Instead, they cling to the male parent as they develop, or they attach themselves to a host species that they parasitize. Even those who remain with the parent, however, have little hope of finding new habitats.

Adults rarely move beyond their immediate surroundings. Some species can swim, but they can neither travel over long distances nor move in a directed manner. Observed in the laboratory, a 1 1/2 inch (38 mm) lentil sea spider is said to be capable of swimming a distance of 4.7 - 5.9 inches (12 to 15 cm) in 30 to 40 seconds. When taken from its habitat and released in the water column, the animal generally extends its legs outward and beats them slowly as it sinks to the bottom. Others tuck their legs in like a closed umbrella and drop quickly to the sea floor.

Lentil sea spider, Crane Neck Point, NY. Long Island Sound

Floating sargassum weeds are said to be responsible for the spread of at least nine species of sea spiders. Other adults have been captured off the bottom clinging to hydroids or other fouling organisms. These, however, were probably torn away from the bottom by trawls or storm waves. Thus it is believed that most species are always close to their favored habitat and food supply.

Pycnogonids are primarily carnivores that often feast on their host. Their diets include the hydroids, anemones, soft corals, sponges, mollusks, polychaete worms and bryozoans (moss animals). Some may feed on algae. Of the marine arthropods, sea spiders are reportedly the greatest predators of moss animals. The mouth of a sea spider is located on the end of a tubular structure known as the proboscis. Some species use the rasping action of their jaws to tear into their food while others may swallow their prey whole. The sea spider *Nymphon rubrum* holds the stem of its favorite hydroid with one of its claws (chelifore) and tears a hole into it with the other. Then, pushing its proboscis firmly up against the hole, it sucks out the tissues. The lentil sea spider uses its claws to remove bits of the stick hydroid *Eudendrium ramosum* and carries the morsels to its mouth. Similar to other species of pycnogonids, little is known of lentil sea spider digestion.

The eyes of lentil sea spiders are their are located on a turret-like structure (ocular tubercle). They are thought to function only in detecting changes in light intensity. Like many other pycnogonids, the lentil sea spider has four eyes lining the sides of the turret, allowing it to obtain information from nearly every direction. ■

The fishes

Naked Goby, *Gobiosoma bosc* (=bosci)

Habitat: Oyster beds, grassy areas, tidal pools and among fouling organisms on dock pilings and man-made refuse.

> Other common names: Clinging goby, variegated goby, common brackish water goby.
> Closely related regional species: Seaboard goby, *G. ginsburgi*.
> Phylum: Chordata. Class: Osteichthyes.
> Order: Perciformes. Family: Gobiidae.
> Geographic range: Connecticut to Cape Kennedy, FL.
>   Pearl Bay, FL, to Campeche, Mexico.
> Salinity tolerance: 0.05 to 45 ppt.
> Maximum size: Large male 2 inches (50 mm), maximum to about 2.5 in. (64 mm) in length.
>   Large female 1.5 inches (37 mm).
> Reproductive season:
>   New York: June through August.
>   Delaware: June through October.
>   Virginia: May through October.
>   Tampa Bay, FL, and Mississippi - April to October.
> Eggs: Unfertilized eggs approximately 0.5 mm in diameter, spherical, yellow and opaque.
> Larvae: Newly hatched larvae 2-2.6 mm in length.
> Life span: Maximum probably not much beyond two years.

*T*he naked goby is a curious, yet skittish little fish whose antics are a delight to watch in captivity or in its natural habitat. Springing off the bottom, the creature swims for just a short distance before settling again to the sea floor. Then, supporting itself on its pelvic fins, it sometimes erects its dorsal fin, arches its back, raises its head and opens its mouth wide as if howling like a coyote. When threatened by another of its kind, the goby approaches the intruder head-on and the pair posture themselves in an exaggerated howling-like stance. In addition, the bodies of one or both fishes can darken in color. The larger of the two occasionally nips or butts its opponent, but the encounter usually ends with no apparent injury to either party. In the confines of an aquarium, the loser avoids the winner.

A confrontation can, however, become more heated. In one case, researchers observed a male seizing a smaller female by the

122

head and shaking her violently before releasing her. The nesting male can be just as feisty.

A male establishes his nest in a dead and gaping oyster shell whose opening is just large enough to permit entry. This allows for a clean and stable substrate and provides protection from preda-

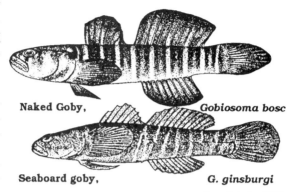

Naked Goby,   *Gobiosoma bosc*

Seaboard goby,   *G. ginsburgi*

From Fritzche, 1978

tors. At times, however, the goby chooses other gaping shells such as the hard-shelled clam, or it excavates a nest beneath an oyster shell. Fanning the sediments with his tail fin and bulldozing the materials with his snout, the fish creates a depression under the shell. He then focuses his attention on a gravid female.

The female is often approached from behind, and once alongside, the male has been observed vibrating his body from head to tail. The movement can propel him sideways. If enticed, the female enters the nest to spawn and then leaves; the male remains to guard their brood. The tiny, round, unfertilized eggs sink and adhere to the floor of the nest and to each other. Soon after fertilization, however, they expand and become elliptical. Development is rapid. Within four to five days, the larvae hatch, leave the nest, and enter the plankton.

The male is a vigilant parent. Throughout the incubation period, he aerates and keeps the eggs clean of silt by fanning them with his tail and pectoral fins. If another fish moves close to the nest, the guardian confronts the intruder and drives it away. In an aquarium, a male who is already protecting a clutch of eggs occasionally allows another gravid female to enter the nest and spawn. As she approaches, however, she can be challenged by the female whose brood is already being guarded. The pair displays the howling-like stance, change color, and may start biting and chasing each other. The male also may become involved. Extending himself partially out of his nest, he opens his mouth and jerks his head up and down as if coughing. He then gently pushes the gravid female with his snout. Though captive males have been observed spawning

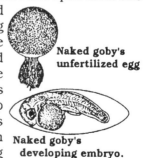

Naked goby's unfertilized egg

Naked goby's developing embryo.

123

Sea grape, *Molgula manhattensis* (a sea squirt).

Naked goby on a dock piling, Groton, CT.

with several females, it is uncertain if they do so in nature. The male remains with his brood until the last of his tiny offspring abandon the nest. At times, however, he leaves his charges to search for food but he always stays close.

Food for the goby consists of small crustaceans, fish, polychaete worms, scavenged material and detritus. Polychaetes, however, often account for over 50 percent of its diet. In an aquarium, the goby feeds ravenously on brine shrimp, and it is assumed that the fish similarly feeds on zooplankton in nature.

Gobies belong to a family of fishes that include more than 700 species worldwide. Most are tropical though a few are found in temperate waters. The naked goby, *Gobiosoma bosc*, and the seaboard goby, *G. ginsburgi*, inhabit temperate coastal Atlantic waters. The naked goby reportedly grows to a length of 2 1/2 inches (64 mm). Near the northern extent of its range, the fish is normally barely more than an inch in length. Its minute size and its secretive nature probably contributed to the assumption by some that it is rare in Long Island Sound (Saltwater Fishes of Connecticut, Bulletin 105, 1971). An 1884 survey of that area, however, revealed that *Gobiosoma* was very common at that time. More recent studies of the Mystic River estuary, CT, and New Haven Harbor, CT, have also shown it to be a resident of the region. Though it has been found in the stomachs of weakfish, bluefish, striped bass, summer flounder and other game fishes, the extent of its importance in the coastal food chain is not yet known.■

Oyster toadfish, *Opsanus tau*

Habitat: Rock reefs on sandy or muddy bottoms.
Oyster reefs, around rocks, aquatic
vegetation or around sunken debris.
Winter habitat: Migrate to deeper water. Also move
to deeper water when water
temperature exceeds 86ºF (30ºC).

Other common names: Toadfish, oyster cracker, sapo.
Phylum: Chordata. Class: Osteichthyes.
Order: Batrachoidiformes. Family: Batrachoididae.
Closely related species: Gulf toadfish, (dog-fish, dog-toad)
Opsanus beta.
Geographic range: Massachusetts to Cape Sable, FL.
Maximum size: 15 inches (381 mm).
Very few longer than 12 in. (305 mm).
Reproductive season: Chesapeake Bay region - April to
July or August.
Woods Hole, MA - May or June to July.
Egg production: Usually fewer than 200 eggs.
Life span: Approximately 8 years in South Carolina.

*T*he oyster toadfish can best be described as having "a face that only a *father* could love." Fleshy tags of tissue hang from its chin, the corners of its mouth and above each eye. Adding to its grotesqueness, the toadfish's scaleless body is covered with a layer of slime. At the risk of giving the fish a human characteristic, it is probably no wonder that the mother abandons her brood. The male, however, guards the nest until the offspring begin to fend for themselves. Though the odd-looking creature will probably never attain commercial value, its ability to produce sounds in excess of 100 decibels - the relative equivalent of a subway train - has made it the subject of a great deal of research. The toadfish, as well as many others, contribute to an often noisy underwater realm.

During the early part of World War II, sonarmen aboard vessels patrolling the warmer portions of the Atlantic and Pacific coasts frequently complained of background noise. Resembling frying bacon, the clamor interfered with the detection of enemy submarines. It had long been known that some fishes were capable of producing sounds, but up to that time, little effort had been made to study them. Sparked by needs of the Navy and the development and improvement of underwater listening devices, research was

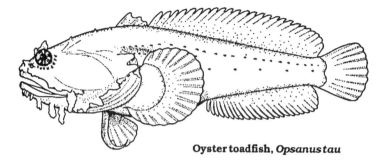

Oyster toadfish, *Opsanus tau*

initiated to identify the sources of biological noise.

Four phyla in the animal kingdom were found to produce significant levels of the underwater sounds. These included the arthropods (crustaceans), mollusks (bivalves), echinoderms (sea urchins) and chordates (whales, seals, fishes).

Snapping shrimps are the best known of the sound-producing arthropods. Also called pistol shrimps, *Alpheus* spp. and *Synalpheus* spp., their sound is apparently emitted when a plunger-like lobe near the base of the creatures' large claw is withdrawn from its socket. The resulting pop has been described as not unlike a cork being pulled from a bottle. Much of the crackling noise detected by the sonarmen was found to be produced by the continuous chorus of millions of snapping shrimps. Knowing that the creatures are restricted to the south temperate and tropical regions, researchers were later surprised to find similar crackling noise on the coast of Rhode Island. Its source was the blue mussel, *Mytilus edulis*. Individual mussels produce the sound by stretching and breaking their anchoring byssal threads as they attempt to reposition themselves (see Blue mussel, page 86). Despite the cacophony of weird noises emitted by a variety of invertebrates, certain fishes inhabiting the shallows along the Atlantic shoreline individually generate louder sounds.

Many of the fishes emit a harsh grating (stridulatory) sound by simply grinding opposing teeth located in their throat (pharyngeal teeth). Ordinarily used to process food, these teeth make a noise when scraped together. Some fishes such as grunts (Pomadasyidae) amplify these sounds with their swim bladder. The northern seahorse, *Hippocampus hudsonius*, also generates a stridulatory sound by rubbing bone against bone; its swim bladder may also act as a resonator.

Sounds are also created by vibrating or squeezing the swim bladder itself. By contracting muscles attached to the outside of, or

126

surrounding the bladder, drums and croakers (Sciaenidae) vibrate the structure and produce a characteristic croak. Using muscles that form the bladder's lateral wall, the oyster toadfish squeezes the heart-shaped or-

> **Invertebrate sound producers:**
> Spiny lobster, *Panulirus* sp.
> Ghost crabs, *Ocypode* sp.
> Fiddler crabs, *Uca* sp.
> Ivory barnacle, *Balanus eburneus*.
> Rock barnacle, *Balanus balanoides*.
> Sea urchin, *Strongylocentrotus droebachiensis*.
> Sea urchin, *Diadema setosum*.

gan and emits a harsh grunt or boatwhistle call. Even with the thin-walled bladder removed from the body, stimulation of its severed nerves generates a sound. Though the male and female swim bladder is similar in structure, it is believed only the male can produce the territorial or mate-attracting boatwhistles. Both sexes can emit an alarm or warning grunt.

In the northern part of its range, the male toadfish establishes a nest from late May to early July. Positioning himself on the underside of a rock or any other convenient site such as a tin can, the male begins to call a mate. A SCUBA diver entering the water during the height of this activity is greeted with a chorus of boatwhistles that creates the impression of being in the middle of a frog pond.

A spawning female readily responds to the boatwhistle. She is even attracted to calls originating from underwater speakers. As she approaches, the summoning male moves aside. She then turns upside down and deposits some 100 eggs on the roof of the nest. The eggs are attached to the surface by means of an adhesive disk. Having fertilized the yellow-colored, 1/5 inch (3 mm) diameter eggs, the male begins his vigil. If he leaves the nest unattended, the ever present cunner, *Tautogolabrus adspersus*, crabs and other predators are quick to devour the entire brood. Guarding the brood, however, is not the male's only role. He also fans the eggs, thereby keeping them aerated and free of silt.

The embryo is barely 1/4 inch (6 to 7 mm) long when it hatches from the top of its egg. Resembling a tadpole with pectoral fins, the creature remains attached as it absorbs the yolk. Unfortunately for the embryo, as it grows it begins to look more and more like its parents. After having reached a length of .63 to .71 inch (16 to 18 mm) the fish starts to feed and breaks free. It then hides under rocks and takes up the sedate life style of an adult.

During the summer,

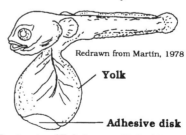

Redrawn from Martin, 1978

Yolk

Adhesive disk

Oyster toadfish larva, .42 in (10.7 mm) in length.

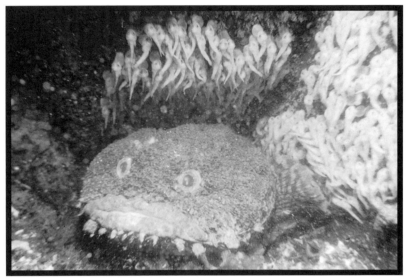

Vigilant father guarding his brood at the nest, Greens Ledge Lighthouse, Darien, CT. Long Island Sound.

the toadfish remains near shore. Occasionally, it is found trapped in tidepools, barely covered with water. The creature has been known to remain alive for up to 24 hours out of water.

Attesting to the toadfish's hardiness, E. W. Gruger, in 1910, reported that even several hours after its partial dissection, the creature snapped at and bit his finger. Prodded with a stick in its underwater habitat, the toadfish snaps at the object and holds it tenaciously for a time. It is doubtful, however, that its short and blunt cone-shaped teeth could penetrate heavy rubber diving gloves. It cannot crack an adult oyster as one of its common names, oyster cracker, seems to imply. When sufficiently threatened by a SCUBA diver, it will, to the dismay of the human observer, often swim directly at him/her instead of the opposite direction. The creature bolts and swims at an amazing speed, no doubt helping it capture a passing prey. With the approach of winter, the toadfish moves offshore where it remains until the spring.■

Tautog (=blackfish), *Tautoga onitis*

Habitat: Rocky bottom, pilings, shipwrecks and other structures.

Other common names: Blackfish, whitechin.
Phylum: Chordata. Class: Osteichthyes.
Order: Perciformes. Family: Labridae.
Closely-related regional species: Cunner (=bergall),
        Tautogolabrus adspersus.
Geographic range: Nova Scotia to South Carolina. Abundant only from
     Cape Cod, MA, to the Delaware Capes.
Maximum size: Approximately 3 feet (.94 m), 22.5 pounds.
    Rare over 14 pounds.
Age at maturity: Males, 3 years of age. Females, 4 years of age.
Adult winter migration: Occurs at approximately 52°F (11°C).
Reproductive season: Massachusetts, June is the principal spawning month.
     Long Island Sound, mid-May to mid-August.
     Maryland and northern Virginia, May to July.
Egg size: 0.89-1.15 mm in diameter.
   Size decreases with increasing water temperature.
Egg production: A 4-year-old female produces about 34,000 eggs;
    a 13-year-old, 457,000 eggs.
     Egg production declines after 16 years of age.
Life span: Males 34 years. Females 22 years.

Steadying itself with a slow beat of its pectoral fins, the tautog hovers just above its prey. Then, with a sudden thrust of its tail, the fish pounces on the crab and bites off one of its claws or legs with an audible crunch. Reeling from the blow, the crustacean tries to escape but it barely recovers before the final onslaught. Within a short time the struggle is over and the predator has its meal.

  The tautog's jaws are equipped with a formidable array of teeth that allow it to consume a variety of hard-shelled creatures. In addition to crabs, the adult fish also feeds on barnacles, hermit crabs, shrimps, lobsters, sand dollars, scallops and 1 to 2 year old blue mussels, *Mytilus edulis*. Mussels are a major part of its diet. When the fish takes hold of one or more of the bivalves with its canine-like teeth, it shakes its head back and forth and tears the creatures' anchoring threads. It then crushes the shells with teeth located in the back of its throat (pharyngeal teeth, see x-ray of a tautog's skull, page 130).

  The juvenile tautog consumes mainly blue mussels and small crustaceans such as amphipods and copepods. In an aquarium, it readily feeds on frozen or live brine shrimp. As might be expected,

129

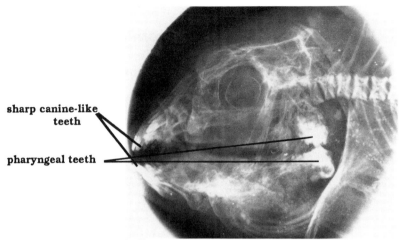

sharp canine-like teeth

pharyngeal teeth

X-ray print of a tautog's skull.

both the adult and juvenile seek out habitats that are close to their food supply and provide them protection from predators.

The tautog remains close to shore throughout the warmer months. The adult inhabits rock reefs, mussel beds, shallow shipwrecks, dock pilings and any other protective structure. A fish that is longer than 12 inches (30.5 cm) in length, generally leaves its shelter during the day and covers a large area as it forages for food. At night, it returns to the same general area where it lies motionless, propped up against some sort of structure. At times, a "sleeping" tautog can be stroked by a SCUBA diver without being disturbed. Once aroused, however, the dazed fish scurries off and occasionally bumps into rocks as it makes its escape. During the day, the fish usually bolts when approached an by intruding diver.

Throughout the season, the juvenile tautog (less than 10 inches, 25 cm) usually ventures no more than a few yards from its shelter. It also lies dormant at night, wedged between, or propped up on, some sort of structure. Some young fishes use sea lettuce (*Ulva lactuca*) or eelgrass (*Zostera marina*) as a nursery habitat. When associated with *Ulva*, they take on a green color that frequently matches that of the seaweed. Most juveniles, however, are light brown in color with blotches of black or brown covering their sides. Older fishes are generally uniformly dark brown or black, and larger specimens often have a white patch on their chins, giving them the name whitechin. Both adult and juvenile respond to seasonal changes in water temperature.

The cool of autumn triggers a migration of the adult to deeper water. In laboratory experiments, the fish enters a dormant state

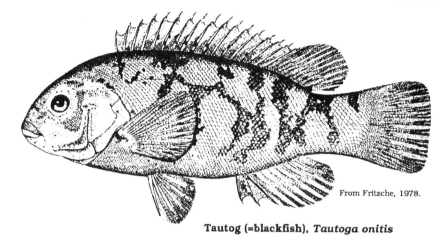

From Fritzche, 1978.

Tautog (=blackfish), *Tautoga onitis*

(torpor) at temperatures of less than 41°F (5°C). During the winter, the juvenile remains at its inshore home site where it can often be found lying on the bottom, wedged between structures, or partially buried under a few millimeters of sand or silt. The fishes remain torpid for about three to four months, and in mid to late spring they become active once more. Soon after returning to the shoreline, the adult begins to show signs of preparation for spawning.

The largest male ordinarily dominantes smaller males and females of his species. When asserting himself, he charges a fish, chases it and sometimes nips at it. Observed in a laboratory, a subordinate male often displays a submissive posture when approached by, or passing near, a dominant male. The behavior consists of the subordinate tilting himself in a head-down position, at an angle of 5° to 90°.

With the onset of the reproductive season, the dominant male becomes more aggressive with other males, and he changes his behavior with respect to the opposite sex. He begins to perform what is termed a "rush" toward a potential mate. In this behavior, the dominant charges the female, veering off within inches of her. The female sometimes reacts to the rush by following the male for a short distance. She also begins to enter his shelter and share his food supply. Occasionally, the female has even been seen taking a clump of mussels from the male's mouth with no apparent reaction from him. During this time the female, whose body is already laden with eggs, is rotund and her color is altered.

The female begins to acquire a mottled white stripe in the middle of each side of her body, a pigmentation referred to as a saddle. Just above her eyes, she also develops white patches known as

131

Spawning behavior: The mating pair swim parallel to each other with the female slightly ahead. As they near the surface, they turn belly-to-belly, break the water's surface and release eggs and milt. From Olla and Samet 1977.

"eyebrows." If the male plays little attention to her, the nuptial shading intensifies, a strategy that may help attract a reluctant male. The changes also become more prominent some 30 to 60 minutes before spawning. The female then raises her dorsal (back) fin, flexes her tail (caudal) fin upward and swims using only her pectoral fins. As she begins her spawning run, the pair swim parallel to each other, toward the surface. Switching to swimming with thrusts of her tail, she propels herself upward and the pair turn their bellies toward each other, forming a U-shape with their bodies. Eggs and milt are released as they break the surface, and the fishes return separately to the bottom. The turbulence created increases the chances of fertilization.

The tiny, transparent eggs are buoyant and concentrate themselves in the upper few meters of water. Eggs spawned early in the season tend to be larger than the ones released during the warmer months. After about two days, the eggs hatch and the 3/32 inch (2.2 mm) long larvae remain attached to their yolk. Then the creatures continues drifting in the plankton until they become bottom dweller at about 25/64 inch (10 mm) in length. ■

Striped bass, *Morone (=Roccus) saxatilis*

Habitat - adults: rock, gravel, sand.
Habitat - juveniles: shallows on clean sandy
or rock bottoms.

Other common names: Striper, rock, rockfish, greenhead,
squidhound, linesider, roller.
Phylum: Chordata. Class: Osteichthyes.
Order: Perciformes. Family: Percichthyidae.
Geographic range: From St. Lawrence River, Canada, to St. Johns River, FL.
Gulf of Mexico from west Florida to Louisiana.
Introduced to Pacific coast in 1879.
Maturity: Males, 2 to 4 years. Females, 3 to 8 years.
Reproductive season: St. Lawrence River, June or July.
Hudson River, mid-May to mid-June.
Chesapeake Bay, April to early June.
North Carolina, late April, May (Fay, 1983).
Cooper River, SC, April to early May.
Savannah River, GA, mid-March to late May.
St. Johns River, FL, mid-February to April.
Peak spawning temperature: Between 63 to 66°F (17 - 19°C).
Eggs: Spherical, semi-buoyant, 1.3 mm in diameter at fertilization.
Egg production: A 4-10 year old female produces about 100,000 mature ova
per year.
Hatching: 48 hours after fertilization at 64°F (18°C).
Size: Maximum recorded weight 125 lbs (56.7 kg).
A striped bass over 30 lbs is likely a female.
Life span: Females up to 29 years or more years; individuals over
12 years old are rare. Fish over 11 years old are usually female.

*T*he autumn-bared trees lining the Hudson River already have begun spreading a new canopy as the striped bass moves upstream to its spawning grounds. The fish is anadromous; it must breed in fresh or nearly fresh water. In the Roanoke River, NC, it travels up to 150 miles from the river's mouth before reaching its destination. The striper, as it is often called, swims approximately 50 to 90 miles upriver in the Potomac. In the Hudson, the majority breed off West Point where the water is essentially fresh. At this site, the current is moderate to strong, a factor believed necessary to insure egg survival and hatching success. If the current is too weak, the

Striped bass, *Morone saxatilis*

semi-buoyant eggs may sink to the bottom, become covered by sediment and suffocate.

The male is first to arrive at the spawning grounds. When an egg-laden female reaches the site, she is soon surrounded by several males. As many as 50 suitors have been observed vying for a position next to a reproductive female. Breaking the surface in a confused melee, the fishes broadcast their eggs and milt. Fertilization takes place in the mixing waters.

A first-time spawner extrudes some 15,000 to 65,000 eggs; an older, larger female can produce a brood of approximately 5 million. For the eggs to survive, a complex relationship must exist between certain physical/chemical (abiotic) factors. These include salinity, temperature, dissolved oxygen, turbidity and current velocity (see table, Physical/chemical factors...). Depending upon temperature, hatching occurs some 29 to 80 hours after fertilization.

The larva, which is .08 -.15 inches (2 to 3.7 mm) long at hatching, carries a supply of yolk on its underside. Its survival depends on physical/chemical factors that are similar to those for the eggs. Just as in the case of the eggs, current is apparently needed to

---

Anadromous fishes* of the Atlantic coast include:

> *These fishes spend most of their lives in the sea and migrate to fresh water to breed.

Sea lamprey, *Petromyzon marinus*: Gulf of St. Lawrence to Florida.

Atlantic sturgeon, *Acipenser oxyrhynchus*: Gulf of St. Lawrence to Gulf of Mexico.

Shortnose sturgeon, *Acipenser brevirostrum*: New Brunswick to Florida.

Atlantic salmon, *Salmo salar*: Greenland to Connecticut.

Brown trout, *Salmo trutta*: Nova Scotia to New Jersey.

Rainbow smelt, *Osmerus mordax*: Gulf of St. Lawrence to New Jersey.

Alewife, *Alosa pseudoharengus*: Gulf of St. Lawrence to Florida.

American shad, *Alosa sapidissima*: St. Lawrence River to Florida.

Blueback herring, *Alosa aestivalis*: Nova Scotia to Florida.

Hickory shad, *Alosa mediocris*: New Brunswick to Florida.

Gizzard shad, *Dorosoma cepedianum*: Massachusetts to Mexico.

Ninespine stickleback, *Pungitius pungitius*: Arctic to New Jersey.

Threespine stickleback, *Gasteros aculeatus*: Hudson Bay to North Carolina.

Atlantic tomcod, *Microgadus tomcod*: S. Labrador to Virginia.

Striped bass, *Morone saxatilis*: St. Lawrence River to Florida.

White perch, *Morone americana*: Nova Scotia to South Carolina.

Possibly anadromous

Opossum pipefish, *Microphis brachyurus*: South Carolina to Brazil.

---

help keep the newly hatched fish suspended off the bottom. A flow that is too strong, however, can wash it from its habitat and expose it to physical/chemical factors that can reduce its chance of survival. An inadequate supply of suitable food can also doom the larva.

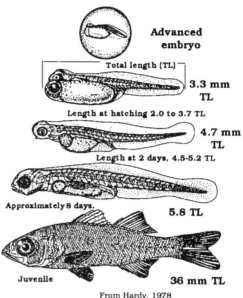

Advanced embryo

Total length (TL)

3.3 mm TL

Length at hatching 2.0 to 3.7 TL

4.7 mm TL

Length at 2 days, 4.5-5.2 TL

Approximately 8 days.

5.8 TL

Juvenile

36 mm TL

From Hardy, 1978

At 2 to 5 days old, the larval fish's mouth is formed, and its supply of yolk is exhausted. It then assumes its role as a predator. The tiny fish is a visual feeder. Spotting a quarry the larva makes a dash toward it. At first the larval fish is rarely successful; it manages to catch its prey only about 2 percent of the time. As it grows, however, it becomes stronger and it consumes progressively larger prey. A larval bass that is less than 25/64 inch (10 mm) long tends to feed mainly on certain larval crustaceans - nauplii; a larva that is over 25/64 inch consumes larger zooplankton.

The larval fish becomes a juvenile at approximately 35 to 50 days of age. The young striper continues to feed on zooplankton and begins to prey on amphipods and fish. In some areas, it consumes mainly mysid shrimps and insect larvae. The growing youngster is much more tolerant of fluctuations in environmental conditions than the eggs and larvae. It gradually moves downriver to higher salinities and remains in the lower part of the river or in nearby estuaries until the approach of its second birthday.

At the age of 2 or 3 years, some of the stripers begin to migrate. North of Cape Hatteras, migrating fish move in a northerly direction in the summer and turn south for the winter. Gulf of Mexico residents, and those from Florida to southern North Carolina, however, do not usually migrate far from their riverine habitat.

Chesapeake Bay and its tributaries account for over 50

Physical/chemical factors in striped bass egg survival and hatching success:
Salinity: optimum, 3-7 ppt.
Temperature: optimum, 64-70°F.
Dissolved oxygen: minimum, 3 to 5 mg/l.
   (Larvae require minimum of 5-6 mg/l).
Current velocity: minimum, 30 cm/sec.
From Fay, 1983, Hill, 1989.

percent of striped bass that migrate along the Atlantic coast. Stripers that are less than a year old usually remain within their natal habitat. As they grow, they join schools numbering from just a few to thousands, and they extend their range

Age/length/sizes:
(length & size approximate)
2 year old/12 in, 3/4 lb
4 year old/20 in, 3 lb.
6 year old/27 in, 10 lb.
8 year old/33 in, 18 lb.

within the bay. Two-year-olds move farther out, but they still tend to limit their range. By age 3, some leave the bay to spend the winter in North Carolina's waters.

Not all of Chesapeake Bay's stripers migrate. During the spring, the migrators move in a northerly direction along the south coast of Long Island. A number of them then continue their trek, reaching as far north as the St. Lawrence River, Canada; about 90 percent of striped bass captured in northern waters are female.

The majority of migrating Hudson River stripers apparently remain within 31 miles (50 km) of the river's mouth (McLaren, 1981). They travel mostly in a northeasterly direction into Long Island Sound with some moving farther up the coast. Sex, age and fish size apparently do not play an important role in the distance traveled by Hudson River stripers. It is also believed that there is little mixing between the Chesapeake Bay and Hudson River stocks, neither during migration nor in the wintering populations.

Historically, striped bass have played an important role in coastal Atlantic fisheries. From the late 1950s through the early 1970s, commercial landings of the fish, between Maine to North Carolina, ranged from 8 million to 14.7 million pounds. Catches by sports fishermen during the same period are thought to have exceeded these numbers. In 1974, however, the harvest began to drop precipitously, reaching a level of 2.4 million pounds by 1982.

Some of the causes for the decline may have included overfishing, destruction or alteration of habitat, environmental contaminants and natural events. A further decline in the harvest to 0.4 million pounds in 1987 was attributed to a management program. To help restore the fishery, states introduced minimum size and creel limits, and, for a time, a moratorium was imposed on commercial fishing.

From FWS Biol Rep. 82(11.118), 1989.

Black sea bass, *Centropristis striata*

Habitat: Adults - rocks, pilings, wharves and
wrecks. Found on patch reefs or inshore
sponge - coral habitats and wrecks
from Cape Fear, NC, to Cape
Canaveral, FL.
Juveniles - appear among inshore
jetties in late May and early June.

Other common names: Atlantic sea bass, blackfish, black bass,
black will, rockfish, tallywag, humpback.
Phylum: Chordata. Class: Osteichthyes.
Order: Perciformes. Family: Serranidae.
Subspecies: *C. striata melana* - Gulf of Mexico.
Two populations of black sea bass may occur along
the Atlantic coast - separated at Cape Hatteras, NC.
Geographic range: Cape Cod, MA, to Cape Canaveral, FL.
Occasionally to Florida Keys.
Depth range: Rare in water deeper than 541 feet (165 m).
Size: To approximately 7.5 pounds.
Northern specimens seldom heavier than 5 pounds.
Egg production: 17,000 to 1,050,000 eggs
Life span: Maximum 10 or more years (Lavenda, 1949).

*M*ost black sea bass begin their lives as females and
later change into males. Though the strategy may
seem odd, such changes in gender from female to male
(protogyny) or from male to female (protandry) are far
from rare in underwater creatures.

Among invertebrates, some 32 species of decapod crustaceans are known to change sex as part of their normal life cycle. The
majority of young eastern oysters, *Crassostrea virginica*, are initially

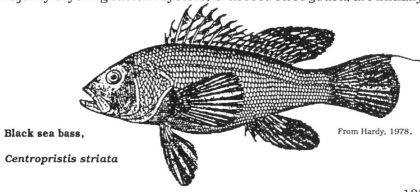

Black sea bass,

Centropristis striata

From Hardy, 1978.

137

male. They change gender around their first birthday. The common European oyster, *Ostrea edulis*, similarly begins life as a male. Once it first changes, however, it then switches back and forth between being female and male. The Eastern melampus (salt-marsh snail) is a simultaneous hermaphrodite. It can function as either sex throughout its reproductive life (see Salt-marsh snail, page 4).

Sex switching fishes include species from at least four different orders. In the wrasse family (Labridae), most if not all of those residing in tropical waters are initially females; males that develop from such females are known as

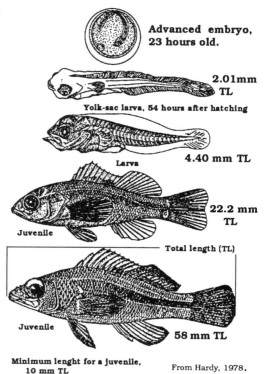

Advanced embryo, 23 hours old.

2.01mm TL

Yolk-sac larva, 54 hours after hatching

Larva   4.40 mm TL

Juvenile   22.2 mm TL

Total length (TL)

Juvenile   58 mm TL

Minimum lenght for a juvenile, 10 mm TL

From Hardy, 1978.

secondary males. The closely-related parrotfishes (Scaridae) are also mainly protogymous - female to male. Some members of these two families, however, hatch as males and remain unchanged throughout their lives; they are called primary males. Black sea bass may also hatch as males and a number of females never make the transition. Those black sea bass that change generally do so between one and five years of age.

Precisely why a female black sea bass switches sex, and what triggers it, is not known. It has been suggested that, in certain other species, the change from female to male may be of advantage when the female's reproductive capacity decreases with age. At times, however, the benefits of change are more obvious. In the Indo-Pacific cleaner fish, *Labroides dimidiatus*, sex change is controlled by a single male. He reigns over a group of three to six females and drives away any male interloper. As long as he dominates, the females in his harem do not change sex. When he dies or is removed from the group, the largest female attempts to take over. If successful in driving away other males, she begins to display the typical male aggressive behavior toward other females within her group. She then assumes

138

the male role more completely and after some two to four days, her transition to male is complete; she (he) then courts and spawns with subordinate females. For those creatures that change from male to female, the advantages also seem more obvious; larger females have the potential for producing a greater number of eggs.

Female sea bass -age at maturity:	
Age	percent of females mature
1	48.4%
2	90.3%
3	99.1%
4 or older	100%
From Mercer, 1989.	

In the majority of black sea bass, sexual succession occurs shortly after spawning. In Florida, the reproductive season begins in February and continues until April. At Woods Hole, MA, breeding has been observed from May to early July. As the spawning period approaches, the male develops a fleshy hump (nuchal hump) on the back of its neck, behind the eyes and in front of the dorsal fin. His body color changes to a bright blue, especially around the eyes and hump.

Eggs and sperm are released offshore and after some 38 to 75 hours, the buoyant eggs hatch and free their tiny larvae. At first, the larval fishes remain offshore (2-51 miles) where they can be found near the surface to a depth of 100 feet (33 m). Larvae longer than 1/2 inch (13 mm) descend toward the bottom and move shoreward. Once along the shore, they take up residence on shell bottoms, eelgrass beds, wharves or jetties.

The developing black sea bass are carnivorous. They prey mainly on small bottom-dwelling crustaceans such as amphipods and isopods. As they grow, they begin to consume shrimps, fishes, mollusks and crabs. Adults prey mainly on crabs and fishes.

The young remain in their estuarine nurseries until the fall when water temperatures drop to about 57°F (14°C). They then migrate to deeper water. The adults follow a similar fall pattern. In the spring, black sea bass from the Middle Atlantic region move north-ward and inshore once again. During this migration, they are actively sought by trawl fishermen. Over the summer, they take up residence in shallow waters, around shipwrecks and rock reefs; in the South Atlantic Bight, they are often found on sponge-coral habitats. In these areas, the fishes are commercially harvested using pots.

Open water visitors

Lion's mane jellyfish, *Cyanea capillata*

Habitat: Adult, coastal waters and open ocean.
Polyp stage attached to shells, rocks, eel
grass and like materials, in rivers and bays.

Other common names: Pink jellyfish, red jellyfish.
Phylum: Cnidaria. Class: Scyphozoa.
Order: Semaeostomeae. Family: Cyaneidae.
Geographic range: Arctic to Florida, Gulf of Mexico.
        Alaska to southern California.
        Worldwide in colder oceans.
Size: Maximum 8 feet across with 200-ft long tentacles.
    Specimens greater than 3 feet in diameter are rare.
Color: Variable - pink, red, yellow, brown.
Reproduction - sexes: Sexes separate, dioecious.
Larvae: Ephyrae - late February in Niantic River, CT.
Occurrence of medusae: Gulf of Maine, April to June.
        Long Island Sound, April to June, and
        August to September.
        Chesapeake Bay, November to early May.
Life span - medusa: Generally short-lived.

*C*alm seas and a mid-day sun occasionally conspire to unveil scores of dark pink, crystalline umbrellas drifting several feet below the surface. Trailing their long slender tentacles, lion's mane jellyfishes, *Cyanea capillata*, maintain their position in the water column with rhythmic pulses of their bell. These beautiful - but to be avoided - creatures inhabit the coast and open ocean from the Arctic to Florida, the Gulf of Mexico, and from Alaska to southern California. Considered part of the zooplankton population, they can be driven into sheltered coves or beaches by strong onshore winds or tidal currents. Once there, the jellyfishes tend to concentrate in certain areas much to the chagrin of bathers and fishrermen. Many of the jellyfishes that have not become stranded, however, are swept away on an outgoing tide.

In Chesapeake Bay, the lion's mane averages 4 to 8 inches across whereas farther north in Long Island Sound, some reach a span of nearly a foot with tentacles exceeding 2 feet (.6 m). Those inhabiting the frigid Arctic waters are known to reach an incredible 8 feet (2.4 m) across, with 200-foot (60 m) tentacles! Specimens greater than 3 feet in diameter, however, are rare.

Jellyfishes belong to phylum Cnidaria, a group that also includes corals, hydroids and anemones (see Northern star coral, page 93). Worldwide, there are approximately 200 known species of jellyfishes, most of which average between .8 to 15.75 inches (2 to 40 cm) in diameter; the lion's mane is the largest of its kind. Among free-swimming species, (stalked jellyfishes, Stauromedusae, are sessile) the jellyfish or medusa represents the adult and sexual stage in their life cycle. For most, sexes are separate.

The first of spring is barely a month off when

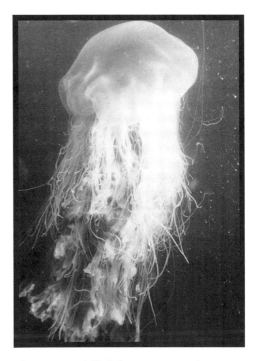

Lion's mane jellyfish, *Cyanea capillata*, Block Island Sound, RI.

larvae of the lion's mane make their appearance in the Niantic River estuary, CT. Known as ephyrae (see illustration, page 142), the snow flake-resembling creatures pulsate constantly. At first, they barely span .039 inch (1 mm) and feed on even smaller zooplankton. As they grow, the young jellyfishes feed on tiny crustaceans such as copepods and amphipods. They also consume mysid shrimps, polychaete worms, fish larvae, larval mollusks and even comb jellyfishes (Ctenophores). By the end of April or early May, the Niantic River jellyfishes have grown into young adults and spawning takes place about three to four weeks later.

In some species of jellyfish, the eggs and sperm are released directly into the water. Those that manage to become fertilized hatch into free-swimming, flat larvae that propel themselves by hairlike cilia. The larvae are known as planulae. Other species of jellyfish such as the lion's mane and the moon jelly, *Aurelia aurita*, brood and protect their eggs and planulae within pits or folds (oral folds) in the oral arms. (The oral arms are ribbon-like structures that extend from the corners of the mouth. They transport food to the mouth. In the

141

lion's mane the arms are folded several times - thus oral folds).

For the Niantic River lion's mane, spawning marks the end of the adult phase. The tentacles are the first to show signs of deterioration followed by the oral folds. With further breakdown of the bell, the creature sinks to the bottom and, at some point, the larval planulae break free. They then attach themselves to rocks, eelgrass, pilings or bivalve shells and metamorphose into polyp larvae known as scyphistomae.

The polyps resemble a hydra. One end is attached to the substrate by a long narrow stalk and the other, the mouth region (oral end), is fringed by tentacles that house stinging cells. Around their base and along short stem-like runners (stolons), the polyps produce yellowish or greenish-brown cysts (podocyst). Each cyst gives rise to another polyp, and, in time, parent polyps and their offspring undergo change (transverse fission) toward their oral end. Known as strobila, these polyps begin producing the ephyrae that grow into adult stingers.

Larval ephyra of the lion's mane. Approximately 2.5 mm in diameter.

The jellyfish's reputation as a stinger played a pivotal role in a Sherlock Holmes mystery, *The Adventure of the Lion's Mane*. The world's greatest detective was called in following the unexplained death of a science teacher. The victim was seen staggering toward the edge of a beach where he collapsed and gasped the words "the lion's mane." Though the jellyfish can produce a severe reaction in sensitive individuals, its toxin does not compare to that of Australia's box jellyfish, *Chironex fleckeri*. The sting of a large box jellyfish is capable of causing death in humans within three to six minutes!

A lion's mane can be equipped with 800 or more tentacles. Throughout their length, the tentacles are covered with batteries of stinging cells (nematocysts) that the creature uses to spear or lasso its prey. Each battery is said to house some two to four dozen nematocysts. Stinging cells also line the edges of the oral lobes. Perhaps due to its small size or differences in the penetration power of its nematocysts, Chesapeake Bay's lion's mane is incapable of stinging most humans. The larger, oceanic variety, however, is much more than a simple nuisance.

> **First aid for lion's mane stings:**
> Unseasoned meat tenderizer or paste of baking soda.
> (Vinegar (5% acetic acid), is recommended for man-of-war and box jellyfish stings, but it may discharge the remaining lion's mane stinging cells; **do not use vinegar with lion's mane stings.**)
> **With any jellyfish sting, do not rinse with fresh water and do not scrub the area or apply ice.**
> (Auerbach, 1991)

For many people, contact with its tentacles elicits an immediate, moderate to sharp burning sensation that lasts about 10 to 20 minutes. The site of the sting becomes red, and the injury can lead to raised welts. Despite its formidable armament, however, the lion's mane plays host to small butterfish (*Peprilus triacanthus*) and other species of juvenile fish.

The young of many species of fish have a natural affinity for nearly anything floating at the surface. They often congregate under a variety of materials that include driftwood, flotsam, jetsam and seaweed. For many of these young and vulnerable creatures, a slow-moving jellyfish offers them open water refuge and a ready

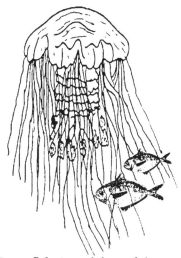

Butterfish: Juvenile butterfish are frequently found swimming in association with the Lion's mane.

supply of food. Apparently unharmed by the stinging tentacles, the juveniles swim in and out of the umbrella's underside. They feed on small marine animals that happen by, or some that already reside on or have become trapped by the jellyfish. As they grow, however, certain species such as the harvestfish, *Peprilus alepidotus*, have been observed feeding directly on their symbiotic partner, the sea nettle *(Chrysaora quinquecirrha)*. Initially, the harvestfish feeds on the tentacles and the underside (manubrium) of its sea nettle host. Later in the season, however, it begins to prey on the bell itself.

A lion's mane and its butterfish symbionts is one of the prettiest and most graceful sights in the coastal Atlantic waters. As the medusa drifts, some of the slaty-blue fishes trail closely behind while others swim among the tentacles. A total of 10 or more butterfish have been observed associated with a single jellyfish. The approach of a diver can prompt the approximately one-inch (30 mm) fishes to dart through the bottom part of the umbrella to the opposite side. If sufficiently frightened, however, the youngsters readily abandon their host. They swim off to another lion's mane or they dive to the relative safety of the depths.■

Studies of Niantic River and Bay, CT, (eastern Long Island Sound) indicate that there may be two different species of *Cyanea*; a river based species and another for the bay. There are morphological (structural) and seasonal differences between these two populations. It is felt that a similar situation exists in the eastern Atlantic (Brewer, 1991).

Longfin squid, *Loligo pealei (=pealeii* AFS 1988)

Habitat: Shoreline and the continental shelf and upper slope. Attaches its eggs to seaweeds, debris and rocks.

Other common names: Long-finned squid, common squid, longfin
      inshore squid, bone squid, winter squid, eyes (small *Loligo*),
      spear squid (Amerika kensakiika, Japanese).
Phylum: Mollusca. Class: Cephalopoda.
Order: Teuthoidea. Family: Loliginidae.
Suborder: Myopsida. About 600 living species of cephalopods.
Geographic range: Nova Scotia to Gulf of Venezuela.
Greatest abundance from Cape Cod, MA, to Cape Hatteras, NC.
Depth range: Shallows to 650 to 1,300 feet (200 to 400 m).
      Rare or absent around islands.
Temperature restrictions: New England, no lower than 46°F (8°C).
      Gulf of Mexico, maximum of 72°F (22°C).
Size: Males, maximum mantle length, 19.7 inches (50 cm).
      Females, maximum mantle length, 15.75 inches (40 cm) .
Reproduction - male: Spermatophores, 8-10 mm long.
Egg production: Approximately 180 per egg capsule.
      A female may produce between 20 to 30 egg capsules.
Life span: Death follows spawning in both sexes.
      They live about one year (Hanlon, 1994).

*L*ongfin squid are sleek and agile hunters. Swimming in schools, the creatures move to the rhythm of their companions as they search for fish, crustaceans or other squid smaller than themselves. They are visual feeders that easily consume 15 percent of their body weight each day. With adequate nutrition, they can grow 0.8 to 1.6 inches (20 to 40 mm) per month.

When prey come into range, hungry squid dart from the school and snatch up one of the creatures with their two tentacles; these appendages can be thrust outward up to 70 percent of their resting length in 15 milliseconds. Held by the tentacles' suction cups, the victim is then passed on to the arms and directed to the mouth. In a final rapid maneuver the squid use their beak to bite their prize behind the head. They go on to devour most of the remaining parts, discarding only the head, tail and intestines of fish prey.

Loligo **capturing a fish.**
    rgb

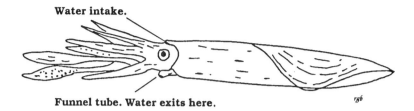

Water intake.

Funnel tube. Water exits here. *rgb*

Squids[1] are mollusks that belong to the class Cephalopoda, a group that also includes cuttlefishes, octopuses and nautili. External shells are only found in the six species of *Nautilus*. Octopuses have no shell at all. Cuttlefishes and squids have greatly reduced shells that are hidden inside the animals. The deep-water cuttlefish (sepioid) *Spirula* has a chambered shell that it carries in the back of its body. When the animal dies the delicate, gas-filled structure floats to the surface and is often found on tropical beaches. Within the soft-bodied longfin squid is a shell that is long and flat. It is known as a gladius or pen. Similar to nautili, octopuses, cuttlefishes and other species of squids, longfin squid swim by rapidly expelling water from their sac-like body (mantle cavity); they are jet-propelled.

A SCUBA diver is fortunate to encounter one of these fascinating cephalopods in their environment. Squids are often seen hovering over the bottom where they sometimes line themselves side-by-side, keeping a watchful eye on an intruder. Any attempt to approach them is usually met with an immediate retreat with the occasional release of a camouflaging cloud of ink. Under the right circumstance, some can jet backward at speeds of up to 25 mph (40 km) per hour. During an escape, flying squids (Onychoteuthis) are known to break the surface and glide for a distance over the water. To propel themselves, squids take in water through vents on each side of their head and force it out through a narrow funnel tube. The openings that allow water to enter the mantle cavity during the filling stage are sealed during the jet stage. All of the exiting water is

gladius or pen

squid *rgb*

Spirula

chambered shell

---

[1] Squid is both singular and plural for the same species.
Squids with the 's' refers to the various species.

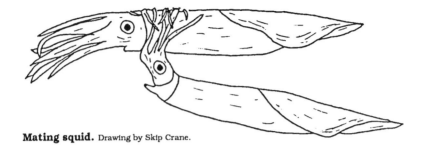

Mating squid. Drawing by Skip Crane.

thus directed through the funnel tube. The animals can move backward, forward, up, down or side to side. They do so by simply pointing the water jet in the direction opposite to the one they want to travel. Squids are also capable of swimming slowly through the water using their fins.

On the New England coast, longfin squid swim inshore annually in response to the season. Their arrival is usually timed for the first week(s) of May and they remain until late November. The winter is spent offshore at depths of 92 to 1,200 feet (28 to 366 m). Larger, sexually mature squid are the first to make their appearance in the spring; these are believed to be 2 years old. They begin to spawn soon after their arrival, and reproductive activities among later arrivals continue through September. South of Cape Hatteras, NC. spawning begins during the late winter. Off Texas, in the Gulf of Mexico, the squid apparently reproduce year-round. Mating behavior, however, has been witnessed mainly in captive squid. Many of the following observations were reported by John Arnold in 1962.

Young longfin squid begin schooling as soon as they can keep up with others of their kind and size. As they grow, they continue to school, paying little attention to the opposite sex. With the approach of spawning, however, mature individuals change their behavior. During this time, the introduction of squid egg capsules in an aquarium are said to trigger reproduction. Males and some females swim to the bottom to explore the egg mass with the tips of their arms. After a period of examination and flushing of the eggs with jets of water, the squid return to the school and the males begin searching for a mate. A successful male swims next to his chosen partner, holding one of his arms up in an S-shaped posture. Dark colored patches form in front of his eyes. If challenged by a competitor, the male raises his display arm even higher and the color patterns become more intense. When a battle ensues, combat is generally confined to a chase or the locking of arms. On rare occasions the battle progresses to bites to the head and fins. The looser then

146

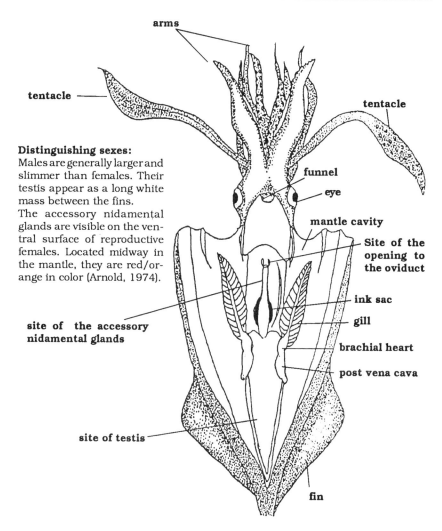

arms

tentacle ———

tentacle

Distinguishing sexes:
Males are generally larger and slimmer than females. Their testis appear as a long white mass between the fins.
The accessory nidamental glands are visible on the ventral surface of reproductive females. Located midway in the mantle, they are red/orange in color (Arnold, 1974).

funnel

eye

mantle cavity

Site of the opening to the oviduct

ink sac

gill

site of the accessory nidamental glands

brachial heart

post vena cava

site of testis

fin

Ventral (underside) of the longfin squid.   Drawing by Skip Crane.

The skin of the squid is only one cell layer thick; it has mucus producing cells that provide protection for its skin. When handled or if it bumps into the walls of an aquarium, it can easily be injured. Bacteria then enter a small wound and weaken the animal. A large wound may cause it to bleed to death, for unlike land vertebrates, squid blood has no clotting mechanism (Hanlon, 1990).

147

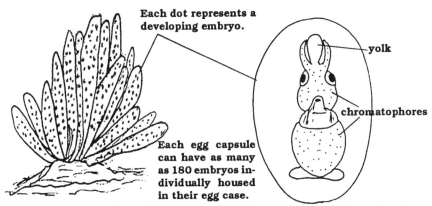

Each dot represents a developing embryo.

yolk

chromatophores

Each egg capsule can have as many as 180 embryos individually housed in their egg case.

Left: Cluster of longfin squid eggs known as a sea mop.

Right: Developing longfin squid embryo within its egg.

retreats to find another female and the winner moves alongside his mate. Holding her with his arms, the male places packets of sperm (spermatophores) in the female's mantle, near the oviduct; mating can also take place face-to-face by locking arms. In both cases, the male transfers the spematophores using a specially adapted arm known as the hectocotylus.

Before depositing her eggs, the female encases them individually in a jelly formed into a finger-like capsule. Some 180 or more eggs are housed in each capsule. As an egg-bearing capsule passes from the oviduct through the funnel, the female gasps it with her arms and holds it with its sticky end outward. She moves to the bottom and generally positions the precious cargo in a cluster of recently deposited capsules. If unavailable, however, she may attach it to an alga such as rockweed (*Fucus* spp.). The female then flushes the cluster with her siphon and swims away to be rejoined by the male. Mating may reoccur and the female often lays as many as 20 to 30 egg capsules. For both sexes, however, the end of spawning marks the end of their lives.

eyes

chromatophores

The chromatophores sometimes appear as large multi-colored patches or as small black dots.

When first deposited, each egg capsule is 3 to 4 inches long (8 to 10 cm) long; they gradually swell to 4.7 to 7 inches (12 to 18 cm). As they develop, the embryos are visible within their pale-amber egg capsule. Resembling miniatures of their parents, the squid hatch within 10 to 27 days. The bodies of these beautiful little

Newly hatched longfin squid.

148

creatures are dotted with multicolored, color producing cells (chromatophores). Most of the 180 or so embryos in each capsule manage to emerge, but thereafter, survival is generally poor. The creatures are food for many zooplankton feeders and the squid must find suitable prey themselves or die of starvation. Despite heavy losses, however, nature provides sufficient numbers to assure survival of the species. ∎

Giant squid, *Architeuthis dux*

Habitat: Probably at or near the bottom, in depths of 3,281 feet+- (1,000 m). Occasionally found washed up on island and coastal beaches.

Species: *Architeuthis dux*, Northern Atlantic.
Architeuthis sanctipauli, Southern Hemisphere.
Architeuthis japonica, Northern Pacific.
(Worldwide, there may be as many as five or more species)
Other names: Frightful polypus, gigantic calamar.
Geographical range: Greenland to the Gulf of Mexico.
Most of the reported sightings occur off Japan and South Africa.
Size: Adults to 60 feet (18 meters), 1,000 pounds (450 kilograms).

*F*act is sometimes difficult to separate from fiction when it comes to giant squid. After returning from voyages that could span several months, some early mariners recounted tales of gargantuan sea monsters. The size of the creatures and intensity of the encounters was probably directly proportional to the time out at sea and the gullibility of the audience.

During the early 1800s, a vessel hoisting anchor off the West African coast reported the sudden attack by a giant squid. "The squid arose and wreathed its fearful snake-like limbs around the vessel's spars" (Buel, 1887). The creatures' tentacles were said to have reached well over the top of the mast. Armed with axes, knives and prayers to their patron St. Thomas, the crew battled the awesome creature and forced it to retreat to the depths. In September 1876, Captain Keene, the master of a coastal Atlantic fishing boat, reported finding a giant squid floating at latitude 44° north and longitude 50° west. "...the body, which was measured as accurately as it could be from a dory, was 50 feet long with tentacles longer than the body. The entire length was more than 100 feet. The tentacles were larger around than a stout man" (Collins, 1884).

From 1871 to 1881, many of these giant cephalopods were discovered on the beaches of Nova Scotia or floating in its offshore waters. There were also an unusual number of giant squid found along the entire coast of the northwest Atlantic during the autumn of 1875.

"Attack of a giant squid".
Redrawn after Buel, 1887. Drawing by Kathy Travers

This prompted well-known zoologist Addison E. Verrill to speculate that their deaths may have been connected with reproductive activity or disease.

In 1980, one of the creatures was stranded on the shores of Plum Island, MA. Measuring some 33 feet (10 m) (body and tentacles), the squid was displayed for a time at Boston's New England Aquarium. Though there have been a few reports of giant squid exceeding a total length of 246 feet (75 m), there is no solid evidence that such an animal exists. The accepted record (total length) stands at 66 feet (20 m) for one that was found on a New Zealand beach in 1880. All of these creatures were found dead or dying. To date, scientists have not had the opportunity to study them in captivity. The Plum Island specimen, however, gave researchers a chance to examine the anatomy of the mysterious giant.

Giant squid have huge eyes that have been described as larger than automobile headlights. "They are the largest eyes in the animal kingdom" (Roper, 1982). In most aspects, giant squid resemble their smaller counterparts, yet in comparison, they are believed to be slow swimmers. Their fins are small, and their body muscles are poorly developed. It is assumed that these giants feed on small fish and other squid, but they probably cannot chase and catch larger, faster-swimming animals. They are themselves an important food item for sperm whales.

Stories of fights-to-the-death between giant squid and sperm whales are legendary. The leviathans have often been found with sucker scars on their skin and around their mouth, giving credence to great clashes between the giants. Sperm whales dive at speeds of nearly 5 mph, to depths that can exceed 3,281 feet (1000 m). Using sonar, they locate and feed on a variety of over 40 species of cephalopod including the giant squid. The remains of giant squid are a commonly found in the gut of these whales. Though the squid's suction cups are thought to be no larger the 2 inches (5.2 cm) in diameter, scars of up to 8 inches (20 cm) have been reported on the skin of sperm whales. The size of the squid was then extrapolated from the size of the scars. Since scars grow as does the whale, their use for determining the size of the squid has led to exaggerated estimates. ■

For further information on these creatures, see Roper, C. and K. Boss. 1982. The giant squid. Scientific American. 246(4):96-105.

Sand tiger shark,
Carcharias (=Odontaspis) taurus
Habitat: Lives near to the bottom.
Remains close to the coast in its
seasonal migrations.

Phylum: Chordata. Class: Elasmobranchiomorphi.
Order: Lamniformes. Family: Ocontaspididae.
Geographic range: Gulf of Maine to Florida and southern Brazil.
                     Mediterranean, tropical West Africa, South Africa,
                     Indo Pacific (close relatives).
Size: Record of 10.5 feet (3.2 m) total length from southwest Florida.
              Specimen caught off Clinton, CT, 8 feet, 10 inches, 250 pounds.
Age of sexual maturity: Male, 4-5 years. Female, approximately 8 years.
Frequency of reproduction: Mature females reproduce annually.
Egg production: Ovary may contain an estimated 22,000 eggs.
Egg size: 0.05 - .4 inch (1.3 - 10 mm)
Offspring - number: 2.
        - size at birth: approximately 32 inches (813 mm).

*S*tewart Springer could not have been prepared for his encounter with the unborn creature. Reaching through an incision in the side of a sand tiger shark, the fisheries scientist explored the uterus. He only wanted to confirm pregnancy in the freshly landed female, but his probing fingers were met by a sudden nip. The tiny shark within was doing what came naturally; it was programmed to consume nearly anything that came within its reach.

Initially, the sand tiger shark, *Carcharias taurus*, produces two eggs, one for each oviduct. The eggs, apparently already fertilized, move a short distance through the oviduct to a specialized gland (oviducal gland) where they are housed in a capsule. They then resume their journey through the oviduct and are deposited each in

151

its own uterus. During this time, the female continues to produce fertilized eggs that play an important role in the development of her embryos.

After some three to four months, the more advanced embryos bite their way through their capsule. Measuring approximately 3.4 inches (60 mm) in length, the creatures are already equipped with an impressive set of teeth. Their role as a predator is postponed, however, while they continue to grow on their remaining supply of yolk and possibly uterine fluids. The first to hatch and/or to reach a length of 3.9 inches (100 mm), move freely in their protected environment. They consume their smaller brothers and sisters and tear up other capsules to get to the unhatched. By the time these roving hunters have reached a length of 9 to 13.4 inches (227 to 340 mm), they have exhausted their supply of siblings. They thus begin to feed on unfertilized, encapsulated eggs that the mother continues to provide (oophagy = egg eating). Each capsule contains seven to 23 eggs. During gestation, a single embryo may consume as many as 17,000 eggs.

The supply of eggs and uterine fluids stretches the sand shark embryos' gut to the size of a soccer ball. The stored nutrients can approach 20 percent of the creatures' body weight. At least one month before birth, the mother ceases egg production. The animals then gradually absorb the stomach contents and their liver increases in size. Following a gestation period of nine to twelve months, the youngsters enter the world head-first; they are about one third the length of their mother. An eight-foot parent can have two offspring that exceed 30-inches in length!

The reproductive strategy seems obvious; a large-sized newborn sand tiger shark has an increased chance of survival. It is believed the embryo's habit of hunting within the uterus makes it a well-trained predator and/or egg eating simply accommodates its rapid rate of growth - the sand tiger shark's embryo grows faster than any known species of elasmobrachs (sharks, skates and rays). Other species of sharks assure the continuance of their kind by producing

Redrawn from Gilmore, 1982.

Unborn sand tiger shark, *Carcharias taurus*:
The creature's gut is filled with eggs and uterine fluids.
Length 31-39.4 in (80 to 100 cm).

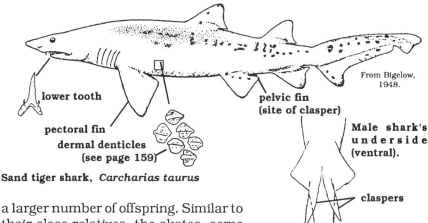

lower tooth

pectoral fin
dermal denticles
(see page 159)

Sand tiger shark, *Carcharias taurus*

pelvic fin
(site of clasper)

From Bigelow,
1948.

Male shark's
u n d e r s i d e
(ventral).

claspers

Male clasper: A male shark's clasper is a modified tubular extension of the pelvic fins that is supported by cartilaginous rods. The rods are often calcified. In some species of sharks such as the sandbar shark, the end of the clasper expands into a fan-like shape when it is inserted into the female's reproductive opening. The male apparently holds himself in the awkward mid-water mating position with his clasper and by biting into the female. Prior to mating, the male flexes his clasper(s) and fills special bladders (siphon sacs) with sea water. Muscles in the walls of the siphon sacs then contract and sea water washes sperm to the female along a groove in the claspers.

a larger number of offspring. Similar to their close relatives, the skates, some of these sharks have eggs that are individually housed in a leathery case (often called a mermaid's purse). The egg cases are attached to algae or any other convenient material, on or near the sea floor, and the developing embryos are abandoned to the mercy of their surroundings. The chain dogfish, *Scyliorhinus retifer*, is said to lay 40 to 70 pairs of encased eggs over an extended period. The sandbar shark, *Carcharhinus plumbeus*, carries its brood internally, giving birth to several youngsters, each approximately 22 to 30 inches long (see Sandbar shark, page 156).

In sharks and other elasmobrachs, the tactic of depositing and abandoning eggs housed in leathery capsules (oviparity = egg laying), is referred to as the most primitive form of reproduction. Yolk attached to the developing embryos serves as food supply until hatching. In other species of sharks, the embryos are housed in a soft egg case, within the mother's uterus (ovoviviparity). The uterus, which develops a highly vascular lining, supplies oxygen to the developing embryo.

In certain species of sharks, only oxygen is supplied. Yolk furnishes the necessary nutrients until birth. For some ovoviviparous species, however, the supply of yolk is not sufficient to last until birth. These creatures (including mako, white [=great white] and

153

sand tiger sharks) thus hatch from their egg case and begin to consume the steady stream of eggs provided by the mother.

Live-bearers (viviparous) are regarded as the most reproductively advanced sharks. As their yolk supply dwindles, the egg case develops a placental attachment to the mother's uterine wall. Nutrition and oxygen are then obtained from the parent and waste products are carried away. The sandbar shark embryo produces such an attachment when it reaches a length of about 12 inches (30 cm). It is viviparous as are most requiem sharks (family Carcharhinidae) and hammerhead sharks (family Sphyrnidae).

Very few people have been lucky enough to observe and record the reproductive behavior of sharks. The chances of witnessing such an event in nature are probably infinitesimal, so scientists have relied on captive sharks. During the late 1980s, Ian Gordon recorded the pre-reproductive and mating behavior of sand tiger sharks. These animals were on display at Oceanworld Manly, in Sidney, Australia. The following are some of his observations.

During the reproductive period, male sand tiger sharks were the first to show changes in their normal routine; they avoided taking food from the surface or from attendant divers. At about the same time, they became more aggressive. The behavior took the form of swimming closer and circling other residents of the tanks. Dominant males followed others of their sex so closely behind that the forward shark's tail movements were restricted; this activity was called tailing (see below). The mature males also inflicted quick physical bites to smaller sharks of species other than their own. Occasionally, these attacks led to fatalities. There was, however, no attempt to eat such victims.

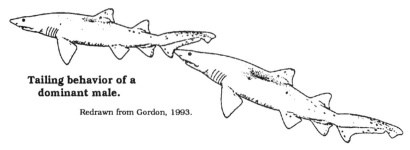

**Tailing behavior of a
dominant male.**

Redrawn from Gordon, 1993.

The sole mature female residing in the tank tended to remain over flat sandy areas. When a suitor approached her, he signaled his interest by biting her on her anal fin and up the flank to the pectoral fin. If the female was not ready to mate, she turned and bit her suitor, driving him off. She then was often seen swimming as close as she could to the bottom, apparently in an attempt to deter any further advances.

As the time to mate approached, the female moved off the bottom and began swimming slowly with the head lower than the tail (submissive position). Her pelvic fins were retracted into a cup shape followed by their

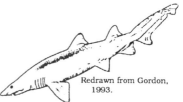

Redrawn from Gordon, 1993.

Female in a submissive position.

flaring. This exposed the opening to her reproductive tract. The dominant male then moved up to the female and bit her on her right flank and the base of the pectoral fin. Swimming slowly by her side, the male turned his body inward to insert his right clasper into the opening of her reproductive tract. Mating lasted from one to two minutes.

Along the Atlantic coast, mature males and females rendezvous each year at the same time and location. Off Florida's east coast, mating takes place in late February to April, at depths of less than 100 feet (30 m). In North Carolina, mating is believed to occur in late April to early May. During the early spring to late summer, pregnant females can be found in the coastal waters south of Cape Hatteras, NC, to Jupiter Island, FL. Others are also found in the northern Gulf of Mexico. Young sand tiger sharks migrate up the Atlantic coast during the summer, taking advantage of the rich estuarine food sources. They then turn southward for the winter. Young males are said to mature at four to five years of age; females mature by their eighth year. ∎

Sand tiger shark, *Carcharias taurus*, in 20 feet of water off Old Field Point Lighthouse, NY. Long Island Sound.

The sand tiger shark feeds mainly on small fishes, including alewives, black drum, bluefish, butterfish, bonito, cunner, tautog, American eel, sea robin, menhanden, mullet, scup, silver hake, black sea bass, various species of flatfishes and small sharks. It also consumes squid, crabs and lobsters. According to the Sears Foundation for Marine Research, there is no record of an unprovoked attack on a human by this shark in North American waters.

155

Sandbar shark,
Carcharhinus (=Eulamia) plumbeus
(=milberti)

Habitat: Females enter shallow bays to give birth.
From shallows along the beaches to 250 m (820 ft).
Young in 5 to 25 fathoms during the summer.
Young in 75 fathoms during the winter off the Carolinas.

Other common names: Brown shark, ground shark.
Phylum: Chordata. Class: Elasmobranchiomorphi.
Order: Lamniformes. Family: Carcharhinidae.
Geographical range: Southern New England to Gulf of Mexico.
                Eastern Caribbean, Mediterranean, Hawaiian Islands,
                South Africa, Indian Ocean and the East China Sea.
Size: Maximum to 9.8 feet (3 m).
          Male matures at 4.3 to 5.8 feet (131 to 178 cm).
          Female matures at 4.7 to 6 feet (144 to 183 cm).
Age at maturity: Up to 30 years (Casey, 1991).
Frequency of reproduction: Every other year.
Reproductive season: Off Salerno, FL, mating takes place during the spring
                  or early summer.
Gestation period: 8 to 12 months; 9 months off Florida.
Offspring - number: 1 to 14; 5 to 12 common.
         - size at birth: 22 to 30 inches (56 to 75 cm).
Life span: May live over 50 years (Casey, 1992).

Man vs Shark

*T*he cry of "shark" on a crowded beach can create panic akin to shouting "fire" in a theater filled to capacity. Unfortunately people's image of these fascinating animals often reflects that of J. W. Buel (1887). The shark's "...eyes are the very personification of cruelty, craftiness and rapacity, being of a greenish cast and peculiarly stony stare. If a man falls into the sea in the presence of this voracious animal, his comrades may at once begin the requiem, or recite prayers for the dead."

People are not a normal food item for any species of sharks. Threatening moves or a bite may represent territorial behavior, or the shark's fear for its own safety. An attack on a person may be a case of mistaken identity such as has occurred with surfers or SCUBA divers. Dressed in a wet suit, they are probably taken for a seal. Divers have been bitten trying to hitch a ride on a shark and spearfishermen have been attacked for their catch. Only about 20 percent of the approximately 350 species of sharks have been implicated in attacks on humans. Of these, more than half belong to the same family, Carcharhinidae or requiem sharks. The white shark (family Lamnidae, mackerel sharks) and the bull shark, tiger shark and oceanic whitetip shark (requiem sharks) are considered the most dangerous species. Sandbar sharks and blacktip sharks are said to be responsible for many of the attacks in Florida waters.

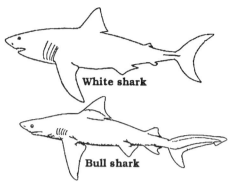

White shark

Bull shark

White shark, *Carcharodon carcharias*. Found in tropical and warm temperate seas, white sharks are rarely over 20 feet in length.

Bull shark, *Carcharhinus leucas*. Found from North Carolina to Brazil and occasionally seen off Long Island, NY. Grows to about 10 feet.

Encounters between these two Florida residents and humans are thought to occur when swimmers splash amongst schools of fish. The approximately 100 yearly shark attacks that do occur worldwide result in about 30 fatalities. These numbers pale, however, in comparison to the human impact on these animals. In 1989, some 4 million sharks were caught in the western North Atlantic; approximately 80 percent of them were killed and dumped back into the sea.

Sharks produce relatively few young, the opposite of many species of bony fishes. For some fishes, a single female can spawn hundreds of thousands of eggs in a single season. But losses due to predation, starvation, disease and pollution, may be high. Nevertheless, the reproductive strategy gives these fishes a tremendous advantage in recovering from natural and man-made (e.g.oil spills, overfishing) disasters. With the right conditions, a small number of mature females can help their population recover within a short time. Sharks with low numbers of offspring have no such advantage. An examination of the sandbar shark's natural history reveals the difficulty faced by this (and other) species in recovering from detrimental human activities.

> **Defaming the good name of a shark:**
> Shark = card shark, cheat, con man, crook,
> fast-talker, fraud, lawyer, swindler.

157

Sandbar Shark

Sandbar sharks produce an average of 8 to 10 youngsters, every other year (see Sand tiger shark, page 151). In preparation for giving birth, pregnant females move into the shallow coastal Atlantic waters, from Long Island, NY, to Cape Canaveral, FL, or the northwestern part of the Gulf of Mexico. During the first five years of their lives, the mid-Atlantic youngsters remain in their near-shore nursery areas for the warm season. Scientists believe the behavior reduces their risk of being eaten by a larger shark. In the cooler season, however, they migrate to sites off the Carolinas, in depths of about 450 feet (137 m). While residing in coastal habitats, the young generally move in the direction of the tidal flow. They feed on blue crab, Atlantic menhaden (*Brevoortia tyrannus*) and other species of crustacean and fishes. Adults tend to consume goosefish (*Lophius americanus*), skates, various flounders, bluefish, weakfish and occasional mollusks and crabs. Though some species of shark ingest trash, the young and adult sandbars are not believed to swallow such indigestible materials.

From Bigelow, 1953

Goosefish, *Lophius americanus*

Sandbar shark, *Carcharhinus plumbeus.* **Block Island Sound, RI.**

158

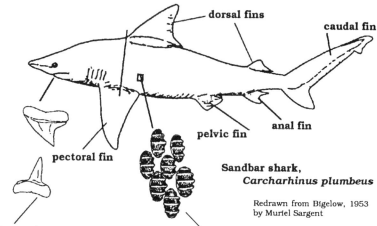

dorsal fins

caudal fin

anal fin

pelvic fin

pectoral fin

Sandbar shark,
Carcharhinus plumbeus

Redrawn from Bigelow, 1953
by Muriel Sargent

The teeth:

Sharks' teeth are very similar to the placoid scales that cover the animals' skin. The teeth are often considered as having developed from scales. The jaw is lined with rows of teeth, one behind the other. As the functional teeth lining the front edge of the jaw wear away or are otherwise lost, they are replaced by teeth lying directly behind them. The size of the replacement teeth are gradually larger, keeping pace with the growth of the animal. The sandbar shark's upper teeth are triangular in shape and sharp, and its lower teeth are long and pointed. The creature is thus well equipped to seize and cut through its prey.

Placoid scales (dermal denticles):

The skin of a shark is covered by abrasive, tooth-like scales known as dermal denticles or placoid scales. The base of each scale is embedded in the skin (dermis). The visible portion, which lies on the surface of the skin, is covered by an enamel called vitreodentine. The ridges on the exposed part of the scales run parallel to the water flow. They apparently decrease drag. As a shark grows, its scales are constantly shed and replaced by larger ones. The sandbar shark's placoid scales are widely spaced; its skin is visible between them.

Male sandbars generally do not move inshore along with pregnant females. Once the females deliver their young, they also move farther offshore. On the Atlantic coast, the adults, like the juveniles, migrate in response to seasonal changes in temperature. In the late fall, they move southward from their mid-Atlantic summering grounds to areas between North Carolina and Florida. Most adult sandbars tracked during a tagging study, traveled some 100 to 1,000 miles between captures. One, however, which was first tagged off Montauk Point, NY, was later retaken 2,028 miles away at Tampico, Mexico. In 1992, the record time between tagging and recapture was 24.9 years. During this interval, the animal had grown only about one inch per year. Studies suggest that the sandbar may take up to 30 years to reach maturity! Since, as previously stated, the shark reproduces only every other year, it should not be surprising that people can seriously affect its population.

During the late 1980s, the author was collecting chiton (small mollusks) in Long Island Sound at a depth of about 15 feet. While rolling over small rocks to inspect their underside, there was a sudden feeling that someone or something was approaching; a 4-foot sand tiger shark came into view. The sun's rays shimmered across the creatures skin as it made a 180° inspection tour of the human that had invaded its space. With a mere flick of its tail, the shark then melted once more into the murky waters of the Sound.

Relatively few of us have the opportunity to encounter one of these magnificent creatures in their environment. All of us, however, may see them in public aquariums. Even there one

Porbeagle shark, *Lamna nasus.*

A tale of over-exploitation: A fishery for the porbeagle shark began in 1961 to satisfy the Italian palate. By 1964, 16 million pounds were harvested. Four years later, only a few hundred of the animals could be found. The population these animals has yet to recover from pre-exploitation levels. (Gruber and Manire, 1990)

Dusky shark, *Carcharhinus obscurus.*

Tiger shark, *Galeocerdo cuvier.*

may be awed by their size, power and beauty. They play an important part in the oceanic food web and fear of them cannot justify their destruction.

From 1974 to 1991, over-exploitation of sandbar, sand tiger, dusky (*Carcharhinus obscurus*) and tiger (*Galeocerdo cuvier*) sharks along the Atlantic coast caused a 60 to 80 percent reduction in their population. "Given the limited ability of sharks to increase their population size, the results (of the report) suggest that stock recovery will probably require decades" (Musick, 1993). ■

The sandbar shark is the most common of the large sharks that enter the bays of Long Island Sound. In very shallow water, it is occasionally seen with its dorsal fin protruding from the surface.

Conversion tables

Metric to U.S. customary.

Length	Abbreviation	=	
1 kilometer	km	=	0.6214 statue miles
1 kilometer	km	=	0.5396 nautical miles
1 meter	m	=	3.280 feet
1 square meter	m²	=	10.76 feet
1 centimeter	cm	=	0.3937 inches
1 millimeter	mm	=	0.03937 inches
1 micron (=1000 mm)	um	=	0.00003937 inches
1 liter	L	=	0.2642 gallons
1 kilogram	kg	=	2.205 pounds
1 gram	g	=	0.03527 ounces
1 milligram	mg	=	0.00003527 ounces
Celsius	°C	=	1.8 X (°C) + 32 = °F

U.S. customary to metric.

1 statue mile	mi	=	1.609 kilometers
1 nautical mile	nmi	=	1.852 kilometer
1 fathom		=	1.829 meters
1 foot	ft	=	.39048 meter
1 inch	in	=	2.54 centimeters
1 inch	in	=	25.40 millimeters
1 gallon	gal	=	3.785 liters
1 pound	lb	=	0.4536 kilograms
1 ounce	oz	=	28.35 grams
1 ounce	oz	=	28350 milligrams
Fahrenheit degrees	°F	=	(°F-32) X .5556 = °C

Others:

1 statue mile	mi	=	5,280 feet
1 nautical mile	nmi	=	6,080 feet
1 statue mile	mi	=	1.15 nautical miles

Salinity: Measured in parts per thousand, ppt or °/₀₀.
Average open ocean salinity, 34.7 ppt.
Dissolved oxygen: Measured in milliliters per liter, ml/L, parts per million, ppm (1 mg O_2/1 liter H_2O = 1 ppm O_2) and % saturation (measured oxygen concentration divided by the theoretical O_2 content at that temperature and conductivity, multiplied by 100 = % oxygen saturation).
Most marine creatures cannot live in water below 3 ppm.
pH (hydrogen ion concentration): Measure of acid/base.
The pH of pure water is 7, neutral pH.
Average oceanic pH, 7.5 to 8.4.

REFERENCES:

Common and scientific names:

Cnidaria:

Cairns, S. et al. 1991. Common and Scientific Names of Aquatic Invertebrates from the United States and Canada: Cnidaria and Ctenophora. Bethesda, MD: American Fisheries Society Special Publ. #22: 75 pp.

Decapod crustaceans:

Williams, A. et al. 1989. Common and Scientific Names of Aquatic Invertebrates from the United States and Canada: Decapod Crustaceans. Bethesda, MD: American Fisheries Society Special Publ. #17: 77 pp.

Fishes:

Robins, C. R. et al. 1991. Common and Scientific Names of Fishes from the United States and Canada. Bethesda, MD: American Fisheries Society Special Publ. #20: 183 pp.

Mollusks:

Turgeon, D. et al. 1988. Common and Scientific Names of Aquatic Invertebrates from the United States and Canada: Mollusks. Bethesda, MD: American Fisheries Society Special Publ. #22: 277 pp.

Species:

American lobster, *Homarus americanus*

Aiken, D. and S. Waddy. 1986. Environmental influence on recruitment of the American lobster, *Homarus americanus*. Can. J. Fish. Aquat. Sci. 43:2258-2270.

Atema, J. and J. Cobb. 1980. Social behavior. In The Biology and Management of Lobsters, Vol I. Cobb, J. and B. Phillips (eds.). New York:Academic Press. p 409-450.

Atema, J. 1986. Review of sexual selection and chemical communication in lobster, *Homarus americanus*. Can. J. Fish. Aquat. Sci. 43:2283-2390.

Cobb, S. J. 1976. The American Lobster: The biology of *Homarus americanus*. Zoology/NOAA Sea Grant, Univ. RI, Mar. Tech. Rep. #49. 32 pp.

Cooper, R. and J. Uzmann. 1971. Migrations and growth of deep-sea lobsters, *Homarus americanus*. Science. 171(3968): 288-290.

Cooper, R. and J. Uzmann. 1980. Ecology of juvenile and adult *Homarus*. In The Biology and Management of Lobsters, Vol II. J. S. Cobb and B. Phillips (eds.). New York:Academic Press. p 97-142.

Eagles, M., D. Aiken and S. Waddy. 1986. Influence of light and food on larval American lobsters, *Homarus americanus*. Can. J. Fish. Aquat. Sci. 43:2303-2310.

Herrick, F. 1911. Natural History of the American lobster. Bull. U.S. Bur. Fish. 29:147-408.

Karnofsky, E., J. Atema, and R. Elgin. 1989a. Field observations of social behavior, shelter use, and foraging in the lobster, *Homarus americanus*. Biol. Bull. 176:239-246.

Karnofsky, E., J. Atema, and R. Elgin. 1989b. Natural dynamics of population structure and habitat use of the lobster, *Homarus americanus*, in a shallow cove. Biol. Bull. 176:247-256.

Lund, W. and Stewart, L. 1970. Abundance and distribution of larval lobsters, *Homarus americanus*, off the coast of southern New England. Proc. Natl. Shellfish. Assn. 60: 40-49.

MacKenzie, C. and J. Moring. 1985. Species profiles: American lobster. U.S. Fish Wildl. Serv. Biol. Rep. 82(11.33). 19 pp.

Smith, E. 1994. Personal communications.

Stewart, L. 1972. The seasonal movements, population dynamics and ecology of the lobster, *Homarus americanus*, off Ram Island, CT. Ph.D. thesis, Univ. CT, Storrs. 112 pp.

Templeman, W. 1934. Mating in the American lobster. Contr. Can. Biol. Fish. 8(32):423-432.

Waddy, S. and D. Aiken. 1991. Mating and insemination in the American lobster, *Homarus americanus*. In Crustacean Sexual Biology, R. Bauer & J Martin (eds.). New York:Columbia Univ. p 126-144.

Waddy, S. 1994. Personal communications.

Weiss, H. M. 1970. The diet and feeding behavior of the lobster, *Homarus americanus*, in Long Island Sound. Ph.D. thesis. Univ. CT., Storrs. 80 pp.

Atlantic oyster drill, *Urosalpinx cinerea*

Costello, D. and C. Henley. 1971. Methods for Obtaining and Handling Marine Eggs and Embryos, 2nd ed. Woods Hole, MA:Marine Biological Laboratory. p 153-154.

Carriker, M.R 1957. Preliminary study of behavior of newly hatched oyster drills, *Urosalpinx cinerea*. 73:328-351.

Carriker, M. and D. Van Zandt, 1972. Behavior of shell-boring muricid gastropods. In Behavior of Marine Animals, H. Winn and B. Olla (eds.). New York:Plenum Press. p 157-242.

Hancock, D. 1956. The structure of the capsule and hatching process in *Urosalpinx cinerea*. Proc. Zool. Soc. London. 127:515-570.

Galtsoff, P. 1964. The American Oyster, *Crassostrea virginica*. Fish. Bull. No. 64, 480 pp.

Ganaros, Anthony, 1958. On development of early stages of *U. cinerea* at constant temperatures and their tolerance to low temps. Biol. Bull 114:188-195.

Griffith, G. and M. Castagna, 1962. Sexual dimorphism in oyster drills of Chicoteague Bay, MD-VA. Chesapeake Sci. 3(3):215-217.

Lebour, Marie V. 1938. The eggs and larvae of the British prosobranchs with special reference to those living in the plankton. J. Mar. Biol Assn. 22:(p 153).

Williams, L., D. Rittschoff, B. Brown, and M. Carriker, 1983. Chemotaxis of oyster drills *Urosalpinx cinerea* to competing prey orders. Biol. Bull. 164:536-548.

Atlantic sand crab (mole crab), *Emerita talpoida*

Caine, E. 1975. Feeding and masticatory structures of selected Anomura (Crustacea). J. Exp. Mar. Biol. Ecol. 18 (3):277-301

Cubit, J. 1969. Behavior and physical factors causing migration and aggregation of the sand crab, *Emerita analoga*. Ecology. 50(1):118-123.

Efford, I. 1966. Feeding in the sand crab, *E. analoga*. Crustaceana 10:167-182.

Jones, L. 1936. A study of the habitat and habits of *Emerita emerita*. Proc. Louisiana Acad. Sci. 3(1):88-91.

Pearse, A., H. Humm and G. Wharton. 1942. A typical sand beach animal, the mole crab, *Emerita talpoida* in ecology of sand beaches at Beaufort, N.C. Ecological Monographs. 12(2):35-190.

Rees, G. 1959. Larval development of the sand crab *Emerita talpoida* in the laboratory. Biol. Bull. 117:356-370

Smith, S. 1877. The early stages of *Hippa talpoida*. Trans. Conn. Acad. Arts. Sci. 3:311-342.

Snograss, R. 1952. The sand crab, *Emerita talpoida* and some of its relatives. Smithsonian Misc. Coll. 117(8):1-34.

Atlantic silverside, *Menidia menidia*.

Austin, H., A. Sosnow, and C. Hickey. 1975. The effects of temperature on the development and survival of eggs and larvae of the Atlantic silverside. Trans. Am. Fish. Soc. 104:762-765.

Conover, D. and B. Kynard. 1981. Environmental sex determination: interaction of temperature and genotype in a fish. Science. 213:577-579.

Conover, D. and S. Murawski. 1982. Offshore winter migration of the Atlantic silverside. Fish. Bull. 80:145-150.

Conover, D. 1992. Personal communications.

Hildebrand, S. and W. Schroeder. 1928. Fishes of Chesapeake Bay. U.S. Bur. Fish. Bull. 53:188.

Holloway, M. 1991. Sex and silversides. Sci. Am. 264(3):24.

Johannes, R. 1978. Reproductive strategies of coastal marine fishes in the tropics. Environ. Biol. Fishes. 3:65-84.

Middaugh, D. 1981. Reproductive ecology and spawning periodicity of the Atlantic silverside. Copeia 1981:766-776.

Middaugh, D., G. Scott, and J. Dean. 1981. Reproductive behavior of the Atlantic silverside. Environ. Biol. Fishes. 6:269-276.

Bay scallop, *Argopecten irradians*

Castagna, M. 1975. Culture of the bay scallop, *Argopecten irradians*, in Virginia. Mar. Fish. Rev. 37(1):19-24.

Fay, C., R. Neves and G. Pardue. 1983. Species profiles: Bay scallop. U.S. Fish and Wildl. Serv. FWS/OBS-82/11.12. 17 pp.

Jones, J. 1986. The puzzling life of the bay scallop. Maritimes. 30(1):14-16.

Land, M. 1966. Activity in the optic nerve of *Pecten maximus* in response to changes in light intensity and to pattern and movement in the optical environment. J. Exp. Biol. 45:83-99.

Moore, J. and E. Trueman. 1971. Swimming of the scallop, *Chlamys opercularis*. J. Exp. Mar. Biol. Ecol. 6:179-185.

Ordzie, C. and G. Garofalo. 1980. Behavioral recognition of molluscan and echinoderm predators by the bay scallop, *Argopecten irradians* at two temperatures. J. Exp. Mar. Biol. Ecol. 43:29-37.

Pohle, D., M. Bricelj and Z. Garcia-Esquivel. 1991. The eelgrass canopy an above-bottom refuge from benthic predators for juvenile scallops, *Argopecten irradians*. Mar. Ecol. Prog. Ser.74:47-59.

Robert, G. 1978. Biological assessment of the bay scallop *Argopecten irradians* for Maritime waters. Can. Fish. Mar. Ser. Rep. No. 778. 13 pp.

Stephens, P. 1978. The sensitivity and control of the scallop mantle edge. J. Exp. Biol.75:203-221.

Tettelbach, S. et al. 1985. A mass mortality of northern bay scallops, *Argopecten irradians*, following a severe spring rainstorm. Veliger 27(4):381-385.

Winter, M. and P. Hamilton. 1985. Factors influencing swimming in bay scallops, *Argopecten irradians*. J. Exp. Mar. Biol. Ecol. 88:227-242.

Black sea bass, *Centropristis striata*

Bigelow, H. and W. Schroeder. 1953. Fishes of the Gulf of Maine. Fish. Bull. 53:407-409.

Hardy, J. 1978. Development of Fishes of the Mid-Atlantic Bight, Vol III. U.S. Fish Wildl. Serv. FWS/OBS-78/12. p 40-49.

Lavenda, N. 1949. Sexual differences and normal protogymous hermaphroditism in the Atlantic sea bass, *Centropristis striata*. Copeia. 1949(3):185-194.

Mercer, L. 1989. Species profiles: Black sea bass. U.S. Fish Wildl. Serv. 82(11.99). 16 pp.

Musick, J. and L. Mercer. 1977. Seasonal distribution of black sea bass, *Centropristis striata*, in the Mid-Atlantic Bight with comments on ecology and fisheries of the species. Trans. Am. Fish. Soc. 106:12-25.

Shapiro, D. 1984. Sex reversal and sociodemographic processes in coral reef fishes. In Fish Reproduction: Strategies and Tactics. G. Potts and R. Wooton (eds.). London: Academic Press. p 103-118.

Black-fingered mud crab (see Say mud crab).

Bloodworm, *Glycera dibranchiata*

Creaser, E. 1973. Reproduction of the bloodworm, *Glycera dibranchiata*, in the Sheepscott Estuary, Maine. J. Fish. Res. Board. Can. 30:161-166.

Dean, D. 1978. The swimming of bloodworms (*Glycera* spp.) at nights, with comments on other species. Mar. Biol. 48:99-104.

Klawe, W. and L. Dickie. 1957. Biology of the bloodworm *Glycera dibranchiata* and its relation to the bloodworm fishery in the Maritime Provinces. Bull. Fish. Res. Board. Can. 115:1-37.

Simpson, M. 1962. Reproduction of the polychaete *Glycera dibranchiata* at Solomons, MD. Biol. Bull 123(2):396-411.

Wilson, W. and E. Ruff. 1988. Species profiles: sand worm and bloodworm. U.S. Fish and Wildl. Serv. Biol Rep. 82(11.80). 23 pp.

Blue crab, *Callinectes sapidus*.

Cameron, J. 1985. Molting in the blue crab. Sci. Am. 252:102-109.

Costlow, J. and C. Bookhout. 1959. The larval development of *Callinectes sapidus* in the laboratory. Biol. Bull. 11:373-396.

Dunham, P. 1978. Sex pheromones in Crustacea. Biol. Rev. 53:555-583.

Gleeson, R. 1991. Intrinsic factors mediating pheromone communication in the blue crab, *Callinectes sapidus*. In Crustacean Sexual Biology, R. Bauer & J. Martin (eds.). New York: Columbia Univ. Press. p 17-32.

Havens, K. and J. McConaugha. 1990. Molting in the mature female blue crab, *Callinectes sapidus*. Bull. Mar. Sci. 46(1):37-47.

Hill, J., D. Fowler and M. Van Den Avyle. 1989. Species profiles: Blue crab. U. S. Fish Wildl. Serv. Biol. Rep. 82(11.100). 18 pp.

Jackowski, R. 1974. Agonistic behavior of the blue crab, *Callinectes sapidus*. Behavior, 50(3-4):232-253.

Tagatz, M. 1968. Biology of the blue crab, *Callinectes sapidus* in the St. Johns River, FL. Fish. Bull. 676:17-33.

Teytaud, A. 1971. The laboratory studies of sex recognition in the blue crab *Callinectes sapidus*. Univ. Miami Sea Grant Tech. Bull. #15. 63 pp.

Spirito, C. 1972. An analysis of swimming behavior in the portunid crab *Callinectes sapidus*. Mar. Behav. and Physiol. 1(3):261-276.

Van Engel, W. 1958. The blue crab and its fishery in the Chesapeake Bay. Part 1. Reproduction, early development, growth, and migration. Commer. Fish. Rev. 20(6):6-17.

Williams, A. 1984. Shrimps, Lobsters and Crabs of the Atlantic Coast of the Eastern United States, Maine to Florida. Washington, DC: Smithsonian Institution Press. p 376-383.

Blue mussel, *Mytilus edulis*

Chanley, P. and J. D. Andrews. 1971. Aids to the identification of bivalve larvae of Virginia. Malacologia. 11(1):45-119.

Field, I. A. 1922. Biology and economic value of the sea mussel, *Mytilus edulis*. Bull. U.S. Bur. Fish. 41:91-179.

Hughes, R. and S, Dunkin. 1984. Behavioral components of prey selection by dogwhelks, *Nucella lapillus*, feeding on mussels, *Mytilus edulis*, in the laboratory. J. Exp. Biol. Ecol. 77:47-68.

Kaplan, S. 1984. The association between the sea anemone *Metridium senile* and the mussel *Mytilus edulis* reduces predation by the starfish *Asterias forbesii*. J. Exp. Biol. Ecol. 79:155-157.

Lee, R. 1975. The structure of a mussel bed and its associated macrofauna. M.S. thesis. Univ. Bridgeport. 56 pp.

Newcombe, C. 1935. A study of the community relationships of the sea mussel, *Mytilus edulis*. Ecology. 16(2):234-243.

Newell, R. 1989. Species profile: Blue mussel. U.S. Fish. Wildl. Serv. Biol. Rep. 82(11). 25 pp.

Petraitis, P. 1987. Immobilization of predatory gastropod, *Nucella lapillus*, by its prey, *Mytilus edulis*. Biol. Bull. 172:307-314.

Theisen, B. 1972. Shell cleaning and deposit feeding in *Mytilus edulis*. Ophelia, 10:49-55.

Bluefish, *Pomatomus saltatrix*

Bigelow, H. and W. Schroeder. 1953. Fishes of the Gulf of Maine. Fish. Bull. 53:383-389.

Hardy, J. 1978. Development of the Fishes of the Mid-Atlantic Bight, Vol lll. U. S. Fish Wildl. Serv. FWS/OBS-78/12. p 340-353.

Lund, W. and G. Maltezos. 1970. Movements and migrations of the bluefish, *Pomatomus saltatrix*, tagged in water of New York and southern New England. Trans. Am. Fish. Soc. 99(4):719-725.

Marks, R. and D. Conover. 1993. Ontogenetic shift in the diet of young-of-the-year bluefish *Pomatomus saltatrix* during the oceanic phase of its early life history. Fish. Bull. 91:97-103.

Norcross, J., S. Richardson, W. Massman, and E. Joseph. 1974. Development of young bluefish, *Pomatomus saltatrix*, and distribution of eggs and young in Virginia coastal waters. Trans. Am. Fish. Soc. 103(3):477-497.

Olla, B., H. Katz, and A. Studholme. 1970. Prey capture and feeding motivation in the bluefish, *Pomatomus saltatrix*. Copeia 1970(2):360-362.

Olla, B., A. Studholme, and A. Bejda. 1985. Behavior of juvenile bluefish, *Pomatomus saltatrix*, in vertical thermal gradients: Influence of season, temperature acclimation and food. Mar. Ecol. Prog. Ser. 23:165-177.

Pottern, G., M. Huish, and J. Kerby. 1989. Species profile: Bluefish. U.S.Fish Wildl. Serv. Biol. Rep. 82(11.94). 20 pp.

Common periwinkle, *Littorina littorea*

Alexander, D. 1960. Directional movements of the intertidal snail, *Littorina littorea*. Biol. Bull. 119(2):301-302.

Brenchley, G. and J. Carlton, 1983. Competitive displacement of the mud snails by introduced periwinkles in the New England intertidal zone. Biol. Bull. 165:543-558.

Clarke, A. and J. Erskine, 1961. Pre-Columbian *Littorina littorea* in Nova Scotia. Science 134:393-394.

Gengron, R. 1977. Habitat selection and migration behavior of the intertidal gastropod *Littorina littorea*. J. Anim. Ecol. 46:79-92.

Gowanloch, J. and F. Hayes, 1926. Contributions to the study of marine gastropods. l. The physical factors, behavior and intertidal life of *Littorina*. Contr. Can. Biol. Fish. 3(4):3-165.

Hadlock, R. 1980. Alarm response of the intertidal snail *Littorina littorea* to predation by the crab *Carcinus maenas*. Biol. Bull. 159:269-279.

Haseman, J. 1911. The rhythmical movements of Littorina littorea synchronous with ocean tides. Biol Bull 21:113-121.

Hayes, F. 1929. Contributions to the study of marine gastropods lll. Development, growth and behavior of *Littorina*. Contr. Can. Biol. Fish. 4(26):41-430.

Lebour, M. 1938. The eggs and larvae of the British prosobranchs with special reference to those living in the plankton. J. Mar. Biol. Assn. 22:126-127.

Moore, H. 1937. The biology of *Littorina littorea* l. Growth of the shell and tissues, spawning, length of life and mortality. J. Mar. Biol. Assn. 21:721-742.

Murphy, D. 1979. A comparative study of freezing tolerances of the marine snails *Littorina littorea* and *Nassarius obsoletus*. Physiol. Zool. 52(2):219-230.

Murphy, D. and L. Johnson. 1980. Physical and temporal factors influencing the freezing tolerance of the marine snail *Littorina littorea*. Biol. Bull. 158:220-232.

Pettitt, C. 1975. A review of predators of *Littorina*, especially those of L. saxatilis. J. Conchol. 28:343-357.

Smith, J. E. and G. E. Newell, 1955. The dynamics of the zonation of the common periwinkle, *Littorina littorea* on a stony beach. J. Animal Ecol. 24:35-56.

Spjelnaes, N. and K. Henningsmoen, 1963. *Littorina littorea*: An indicator of Norse settlement in North America? Science 141:275-276.

Wells, H. W., 1965. Maryland records of the gastropod *Littorina littorea*, with a discussion of factors controlling its southern distribution. Chesapeake Sci. 6(1):38-42.

Common sea star, *Asterias forbesi*

Aldrich, J. 1976. The spider crab *Libinia emarginata* and the starfish, an unsuitable predator but a cooperative prey. Crustaceana 31(2):151-156.

Burnett, A. 1960. The mechanism employed by the starfish *Asterias forbesi* to gain access to the interior of the bivalve *Venus mercenaria*. Ecology 41:583-584.

Campbell, D. 1984. Foraging movements of the sea star *Asterias forbesi* in Narragansett Bay, RI. Mar. Behav. Physiol. 11:185-198.

Coe, W. 1972. Starfishes, Serpent Stars, Sea Urchins and Sea Cucumbers of the Northeast. New York: Dover. 152 pp.

Doering, P. 1981. Observations on the behavior of *Asterias forbesi* feeding on *Mercenaria mercenaria*. Ophelia 20(2):169-177.

Doering, P. 1982. Reduction of attractiveness to the sea star *Asterias forbesi* by the clam *Mercenaria mercenaria*. J. Exp. Mar. Biol. Ecol. 60:47-61.

Doering, P. 1982. Reduction of sea star predation by burrowing response of the hard clam *Mercenaria mercenaria*. Estuaries 5(4):310-315.

Kaplan, S. 1984. The association between the sea anemone *Metridium senile* and the mussel *Mytilus edulis* reduces predation by the starfish *Asterias forbesii*. J. Exp. Mar. Biol. Ecol. 79:155-177.

Lavoie, M. 1956. How sea stars open bivalves. Biol. Bull. 111:114-122.

Loosanoff, V. 1961. Biology and methods of controlling the starfish, *Asterias forbesi*. U.S. Dept. Interior, Fishery Leaflet #520.

Loosanoff, V. 1964. Variations in time and intensity of setting of the starfish, *Asterias forbesi*, in Long Island Sound during a twenty-five year period. Biol. Bull. 126:423-439.

MacKenzie, C. 1970. Feeding rates of the starfish *Asterias forbesi*, at controlled water temperatures and during different seasons of the year. Fish Bull. 68:67-72.

Mackenzie, C. 1977. Predation of the hard clam, *Mercenaria mercenaria*, populations. Trans. Am. Fish. Soc. 106:530-537.

Mackenzie, C. 1992. Personal communications.

Prior, D., A. Schneiderman, and S. Greene, 1978. Size dependent variation in the evasive behavior of the bivalve mollusc *Spisula solidissima*. J. Exp. Biol. 78:59-75.

Smith, I. 1961. Clam-digging behavior of the starfish, *Pisaster brevispinus* on soft substrate. Behavior 18:148-153.

Diamondback terrapin, *Malaclemys terrapin*
Allen, J. and R. Littleford. 1955. Observations on the feeding habits and growth of immature diamondback terrapins. Herpetologica 11(1):77-80.
Burger, J. 1976. Behavior of the hatchling diamondback terrapins, *Malaclemys terrapin*, in the field. Copeia 1976:742-748.
Burger, J. and W. Montevecchi. 1975. Nest site selection in the terrapin, *Malaclemys terrapin*. Copeia 1975:113-119.
Daiber, F. 1982. Animals of the Tidal Marsh. New York:Van Nostrand Reinhold. 422 pp.
Ernst, C. and R. Barbour. 1972. Turtles of the United States. Lexington: Univ. Kentucky Press. 347 pp.
Finneran, L. 1948. Diamondback terrapin in Connecticut. Copeia 1948(2):138.
Hildebrand, S. 1929. Review of experiments on artificial culture of diamondback terrapin. U.S. Fish. Bull. 45:25-70.
Hildebrand, S. 1932. Growth of the diamond-back terrapins, size attained, sex ratio, and longevity. Zoologica 9(15):552-563.
Hurd, L., G. Smedes, and T. Dean. 1979. An ecological study of a natural population of diamondback terrapins, *Malaclemys t. terrapin*, in a Delaware salt marsh. Estuaries 2(1):28-33.
Montevecchi, W. and J. Burger. 1975. Aspects of the reproductive biology of the northern diamondback terrapin, *Malaclemys t. terrapin*. Am. Midl. Nat. 94(1):166-178.
Palmer, W. and C. Cordes. 1988. Habitat suitability index models: Diamondback terrapin, Atlantic coast. U.S. Fish Wildl. Serv. Biol. Rep. 82(10.151). 18 pp.
Yearicks, E. R. Wood, and W. Johnson. 1981. Hibernation of the northern diamondback terrapin, *Malaclemys t. terrapin*. Estuaries 4(1):78-80.

Eastern melampus (see Salt-marsh snail)

Eastern mud snail, *Ilyanassa obsoleta*
Atema, J. and G. Burd. 1975. A field study of chemotactic responses in the marine mud snail, *Nassarius obsoletus*. J. Chem. Ecol. 1(2):243-251.
Atema, J and D. Stenzler, 1966. Alarm substance of the marine mud snail, *Nassarius obsoletus*: biological characterization and possible evolution. J. Chem Ecol. 3(2)173-187.
Carr, W. 1967. Chemoreception in the mud snail, *Nassarius obsoletus*. Properties of stimulatory substances extracted from shrimp. Biol. Bull. 133:90-105.
Crisp, M. 1969. Studies on the behavior of *Nassarius obsoletus*. Biol. Bull. 156:355-373.
Pechenik, J. 1975. The escape of veligers from the egg capsules of *Nassarius obsoletus* and *Nassarius trivittatus*. Biol. Bull. 149:580-589.
Scheltema, R. 1962. Pelagic larvae of the New England intertidal gastropods, *Nassarius obsoletus* and *N. vibex*. Am. Microscopical Soc. 81:1-11.
Scheltema, Rudolf S. 1964. Feeding habits and growth in the mud snail *Nassarius obsoletus*. Chesapeake Sci. 5:161-166.
Stenzler, D. and J. Atema. 1977. Alarm response of the mud snail *Nassarius obsoletus*: specificity and behavioral priority. J. Chem. Ecol. 3(2):159-171.
Sullivan, C. and T. K. Maugel. 1984. Formation, organization and composition of the egg capsule of the marine gastropod, *Ilyanassa obsoleta*. Biol. Bull. 167:378-389.
Sullivan, C. and D. Bonar. 1985. Hatching of *Ilyanassa obsoleta* embryos. Biol. Bull. 169:365-376.

Grass shrimp, *Palaemonetes* spp.
Broad, A. 1957. Larval development of *Palaemonetes pugio*. Biol. Bull. 112:144-161.
Burkenroad, M. D. 1947. Reproductive activities of decapod crustacea. Am. Naturalist.

81(800):392-398.

Daiber, F. 1982. Animals of the Tidal Marsh. New York:Van Norstrand Reinhold. 422 pp.

Hoffman, C. 1980. Growth and reproduction of *Palaemonetes pugio* and *P. vulgaris* populations in Canary Creek Marsh, Delaware. MS thesis, Univ. DE. 125 pp.

Jenner, C. 1955. A field character for distinguishing *Palaemonetes vulgaris* from *P. pugio*. Biol. Bull. 109:360.

Welsh, B. 1975. The role of grass shrimp, *Palaemonetes pugio* in a tidal marsh ecosystem. Ecology 56:513-530.

Williams, A. B. 1984. Shrimps, Lobsters, and Crabs of the Atlantic Coast of the Eastern United States, Maine to Florida. Smithsonian, Washington, D.C. p 71-78.

Green crab, *Carcinus maenas*

Broekhuysen, G. 1936. On development, growth and distribution of *Carcinus maenas*. Arch. Gnarl. Zool. 2:257-399.

Crothers, J. 1967. The biology of the shore crab, *Carcinus maenas*. The background-anatomy, growth and life history. Fld. Stud. 2:407-434.

Crothers, J. H. 1968. The biology of the shore crab, *Carcinus maenas*. The life of the adult crab. Fld. Stud. 2:579-614.

Edwards, R. L. 1958. Movements of individual members in a population of the shore crab, *Carcinus maenas* in the littoral zone. J. Anim. Ecol. 27:37-45.

Foxon, G. 1939. Notes on the history of *Sacculina carcini*. J. Mar. Biol. Assn. UK. 24:253-264.

Glude, J. 1955. The effects of temperature and predators on the abundance of the soft-shell clam, *Mya arenaria* in New England. Trans. Am. Fish. Soc. 84:13-26.

Naylor, E. et al, 1971. External factors influencing the tidal rhythm of shore crabs. Proc. 2nd Int. Interdisc. Conf. Res. 2:173-180.

Rasmussen, E. 1959. Behavior of sacculinized shore crabs (*Carcinus maenas*). Nature, 183(4659):479-480.

Ropes, J. 1968. The feeding habits of the green crab, *Carcinus maenas*. Fish. Bull. 67(2): 183-203.

Welch, W. 1968. Changes in the abundance of the green crab, *Carcinus maenas* in relation to recent temperature changes. Fish. Bull. 67(2): 337-345.

Horseshoe crab, *Limulus polyphemus*

Barlow, R. et al. 1986. Migration of *Limulus* for mating: relation to lunar phase, tide height, and sunlight. Biol. Bull., 171:310-329.

Botton, R. and R. Loveland. 1987. Orientation of the horseshoe crab, *Limulus polyphemus*, on a sandy beach. Biol. Bull. 173:289-298.

French, K. 1979. Laboratory culture of embryonic and juvenile *Limulus*. In Biomedical Applications of the Horseshoe Crab. E. Cohen (ed.). New York:A. R. Liss. p 61-71.

Novitsky, T. 1991. The blood of the horseshoe crab. Oceanus 27(1):13-18.

Pearl, R. On the behavior and reactions of *Limulus* in the early stages of its development. J. Comp. Neurology and Psychology. 14:138-164.

Pomerat, C. 1933. Mating in *Limulus polyphemus*. Biol. Bull. 64:243-252.

Ropes, S. 1961. Longevity of the horseshoe crab, *Limulus polyphemus*. Trans. Am. Fish. Soc. 90(1):79-80.

Rudloe, A. 1979. *Limulus polyphemus*: A review of the ecologically significant literature. In Biomedical Applications of the Horseshoe Crab. E. Cohen (ed.). New York:A. R. Liss. p 27-35.

Rudloe, A. 1980. The breeding behavior and pattern of movement of the horseshoe crab, *Limulus polyphemus*, in the vicinity of breeding beaches in Apalachee Bay, FL. Estuaries, 3(3):177-183.

Sekiguchi, K. and K. Nakamura. Ecology of the extant horseshoe crab. In Biomedical Applications of the Horseshoe Crab. E. Cohen (ed.). New York:A. R. Liss. p 37-45.

Shuster, C. 1979. Biology of *Limulus*. In Biomedical Applications of the Horseshoe Crab. E. Cohen(ed.). New York:A. R. Liss. p 1-26.

Shuster, C. 1982. A pictorial review of the natural history and ecology of the horseshoe crab *Limulus polyphemus*, with reference to other *Limulidae*. In Physiology and Biology of the Horseshoe Crab. J. Bonaventura, et al (eds.). New York:A. R. Liss. p 1-52.

Lentil sea spider, *Anoplodactylus lentus*

Arnaud, F. and R. Bamber. 1987. The Biology of Pycnogonida. In Advances in Marine Biology, J. Blaxter and A. Southward (eds.). New York: Academic Press. p 1-96.

Cole, L. 1901. Notes on the habits of pycnogonids. Biol. Bull. 2(5):195-207.

Cole, L. 1906. Feeding habits of the pycnogonid, *Anoplodactylus lentus*. Zool. Anz. 29:740-741.

King, P. 1973. Pycnogonids. New York:St. Martins. 144 pp.

Ryland, J. 1976. Pycnogonid predators. In Physiology and Ecology of Marine Bryozoans. Advances in Marine Biology. New York: Academic Press. 14:417-421.

Lion's mane jellyfish, *Cyanea capillata*

Auerbach, P. 1991. A Medical Guide to Hazardous Marine Life. St. Louis:Mosby. p 27.

Brewer, R. 1989. The annual pattern of feeding, growth and sexual reproduction of *Cyanea* in the Niantic River estuary, Connecticut. Biol. Bull. 176:272-281.

Brewer, R. 1991. Morphological differences between, and reproduction isolation of, two populations of the jellyfish *Cyanea* in Long Island Sound. Hydrobiologica. 216/217:471-477.

Berrill, N. 1949. Form and growth in the development of a scyphomedusa. Biol. Bull. 96:283-292.

Burnett, J. and G. Calton. 1987. Venomous pelagic coelenterates: chemistry, toxicology, immunology and treatment of their stings. Toxicon. 25(6):581-602.

Dahl, E. 1961. The association between young whiting, *Gadus merlangus*, and the jellyfish *Cyanea capillata*. Sarsia. 3:47-55.

Grondahl, F. and L. Hernroth. 1987. Release and growth of *Cyanea capillata* ephyrae in the Gullmar Fjord, western Sweden. J. Exp. Mar. Biol. Ecol. 106:91-101.

Rice, N. and W. A. Powell. 1972. Observations on three species of jellyfishes from Chesapeake Bay with special reference to their toxins. 11 *Cyanea capillata*. Biol. Bull. 143:617-622.

Longfin squid, *Loligo pealei*, and Giant Squid

Arnold, J. 1962. Mating behavior and social structure in *Loligo pealii*. Biol.Bull. 123:53-57.

Arnold, J. 1990. Squid mating behavior. In Squid as Experimental Animals, D. Gilbert, W. Adelman and J. Arnold (eds.). New York:Plemum. p 65-75.

Barnes, R. Invertebrate Zoology. Philadelphia:Saunders. p 442-465.

Buel, J. 1887. Sea and Land. Philadelphia:Historical Publ. p 68-84.

Collins, J. 1884. A large squid. Bull. U.S. Fish Comm. 4:15.

Frost, N. 1934. Notes on a giant squid captured at Dildo, Newfoundland. Rep. Newfoundland Fish. Res. Comm. 2(2):100-113.

Gosline, J. and M. E. DeMont. 1985. Jet-propelled swimming in squids. Sci. Am. 252(1):96-103.

Hanlon, R. 1990. Maintenance, rearing and culture of Teuthoid and Sepioid squids. In Squid as Experimental Animals, D. Gilbert, W. Adelman and J. Arnold (eds.).

New York:Plemum. p 35-61.

Hanlon, R. 1994. Personal communications.

Roper, C. and K. Boss. 1982. The giant squid. Sci. Am. 246(4):96-105.

Summers, W. 1983. *Loligo pealei*. In P. Boyle (ed.) Cephalopod Life Cycles. New York:Academic Press, p 115-142.

Wells, M. and R. O'Dor. 1991. Jet propulsion and the evolution of the cephalopods. Bull. Mar. Sci. 49(1-2):419-432.

Longwrist hermit crab, *Pagurus longicarpus*

Bachand, R. Personal observations.

Bertness, M. 1981. Influence of shell-type on hermit crab growth rate and clutch size. Crustaceana 40(2):197-205.

Blackstone, N. 1984. The effects of history on the shell preference of the hermit crab *Pagurus longicarpus*. J. Exp. Biol. Ecol. 81:225-234.

Blackstone, N. 1985. The effects of shell size and shape on the growth and form in the hermit crab *Pagurus longicarpus*. Biol. Bull. 168:75-90.

Christensen, H. 1967. Ecology of *Hydractinia echinata*. Ophelia 4:245-275.

Conover, M. 1976. The influence of some symbionts on the shell-selection behavior of the hermit crabs, *Pagurus pollicaris* and *Pagurus longicarpus*. Animal Behavior 24:191-194.

Elmwood, R. and S. Neil, 1992. Assessment and Decisions: A study of information gathering by hermit crabs. New York:Chapman and Hall. p 37-44.

Fink, H. 1940. Deconditioning of the fright reflex in the hermit crab *Pagurus longicarpus*. J. Comp. Psychol. 32(1):33-39.

Hatfield, P. 1965. *Polydora commensalis*. Larval development and observations on adults. Biol. Bull. 128:356-368.

Hazlett, B. 1972. Ritualization of marine crustacea. In Behavior of Marine Animals, Vol 1, H. Winn and B. Olla (eds.). New York: Plenum Press. p 97-125.

Hazlett, P. 1992. Personal communications.

Johnson, R. and J. Ebersole, 1989. Seasonality in the reproduction of the hermit crab *Pagurus longicarpus*. Crustaceana 57(3):311-313.

Mercado, N. and C. Lytle 1980. Specificity in the association between *Hydractinia echinata* and sympatric species of hermit crabs. Biol. Bull. 159:337-348.

Rebach, S. and D. Dunham. 1974. Studies in Adaptation. New York:John Wiley. p 217-264.

Reese, E. 1962. Shell selection behavior of hermit crabs. Animal Behavior 10(3-4):347-230.

Reese, E. 1963. The behavioral mechanism underlying shell selection by hermit crabs. Behavior 21:78-124.

Roberts, Morris H. 1970. Larval development of *Pagurus longicarpus* reared in the laboratory. Biol. Bull. 139:188-202.

Scully, E. 1978. Utilization of surface foam as a food source by the hermit crab, *Pagurus longicarpus*. Mar. Behav. Physiol. 5:159-162.

Scully, E. 1979. The effects of gastropod shell availability and habitat characteristics on shell utilization by the intertidal hermit crab *Pagurus longicarpus*. J. Exp. Mar. Biol. Ecol. 37:139-152.

Scully, E. P. 1992. Personal communication.

Spight, Tom M. 1977. Availability and use of shells by intertidal hermit crabs. Biol. Bull. 152:120-133.

Mantis shrimp, *Squilla empusa*

Caldwell, R. 1987. Assessment strategies in stomatopods. Bull. Mar. Sci. 41(2):135-150.

Caldwell, R. 1991. Variation in reproductive behavior in stomatopod Crustacea. In Crustacean Sexual Biology, R. Bauer and J. Martin (eds.),. New York: Columbia

Univ. Press. p 67-90.

Caldwell, R. 1993. Personal communications.

Caldwell, R. and H. Dingle. 1976. Stomatopods. Sci. Am. 234:80-89.

Fotheringham, N. and S. Brunenmeister, 1989. Beachomber's Guide to Gulf Coast Marine Life. Houston: Gulf Publ. p 47-48.

McCluskey, W. 1977. Surface swarming of *Squilla empusa* in Narragansett Bay, RI. Crustaceana. 33(1):102-103.

Morgan, S. 1980. Aspects of larval ecology of *Squilla empusa* in Chesapeake Bay. Fish. Bull. 78(3):693-700.

Myers, A. 1979. Summer and winter burrows of a mantis shrimp, *Squilla empusa*, in Narragansett Bay, RI. Estuarine & Coastal Mar. Sci. 8:87-98.

Mummichog, *Fundulus heteroclitus*

Able, K. and M. Castagna. 1975. Aspects of an undescribed reproductive behavior in *Fundulus heteroclitus* from Virginia. Chesapeake Sci. 16(4):282-284.

Abraham, B. 1985. Species profile: Mummichog and striped killifish. U.S. Fish Wildl. Serv. Biol. Rep. 82(11.40). 23 pp.

Armstrong, P. and J. Child. 1965. Stages in the normal development of *Fundulus heteroclitus*. Biol. Bull. 128(2):143-168.

Chidester, F. 1920. The behavior of *Fundulus heteroclitus* on the salt marshes of New Jersey. Am. Nat. 54:551-557.

DiMichele, L., M. Taylor, and R. Singleton. 1981. The hatching enzyme of *Fundulus heteroclitus*. J. Exp. Zool. 216(1):133-140.

Hoffman, R.,et al. 1978. Effect of prehatching weightlessness on adult fish behavior in dynamic environments. Aviat. Space Environ. Med. 49(4:576-581.

Keenleyside, M. 1979. Diversity and adaptations in fish behavior. New York:Springer-Verlag. 208 pp.

Newman, H. 1907. Spawning behavior and sexual dimorphism in *Fundulus heteroclitus* and allied fish. Biol. Bull. 12:314-345.

Taylor, M. et al. 1977. Lunar spawning cycle in the mummichog, *Fundulus heteroclitus*. Copeia. 1977(2):291-297.

Taylor, M., L. DiMichele, and G. Leach. 1977. Egg stranding in the life cycle of the mummichog, *Fundulus heteroclitus*. Copeia 1977(2):397-399.

Naked goby, *Gobiosoma bosci*

Dahlberg, M. and J. Conyers, 1973. An ecological study of *Gobiosoma bosci* and *G. ginsburgi* on the Georgia coast. Fish. Bull. 71(1)279-287.

Fritzsche, R. 1978. Development of the Fishes of the Mid-Atlantic Bight. Vol V. U. S. Department of the Interior, FWS/OBS 78-/12. p 214-220.

Hilderbrand, S. and L. Cable, 1938. Further notes on the development and life history of some teleosts at Beaufort, N.C. Bull. U.S. Bur. Fish. 48:505-642.

Hilderbrand, S. and W. Schroeder, 1928. Fishes of Chesapeake Bay. Bull. U.S. Bur. Fish. 43(1):1-366.

Nero, L. 1976. The natural history of the naked goby *Gobiosoma bosci*. MS thesis, Old Dominion Univ. 85 pp.

Northern rock barnacle, *Semibalanus (=Balanus) balanoides*

Barnes, H. and M. Barnes. 1959. A comparison of the annual growth patterns of *Balanus balanoides* with particular reference to the effect of food and temperature. Oikos 10:1-18.

Barnes, H. and M. Barnes. 1968. Egg numbers, metabolic efficiency of egg production and fecundity: Local and regional variations in a number of common cirripeds. J. Exp. Mar. Biol. Ecol. 2:135-153.

Barnes, R. 1987. Invertebrate Zoology, 5th ed. Philadelphia: Saunders. p 580-586.

Linder, E. 1984. The attachment of macrofouling invertebrates. In Marine Biodeterioration: An Interdiciplinary Study. J. Costlow and R. Tipper (eds.) Annapolis: Naval Institute Press. 384 pp.

Stubbings, H. 1975. *Balanus balanoides*. Memoir No 37. Liverpool: Liverpool Univ. Press. p 17-170.

Northern searobin, *Prionotus carolinus*

Bigelow, H and W. Schroeder. 1953. Fishes of the Gulf of Maine. Fish. Bull.53:467-472.

Finger, T. 1982. Somatotopy in the representation of the pectoral fin and free fin rays in the spinal cord of the sea robin, *Prionotus carolinus*. Biol. Bull. 163:154-161.

Fish, M. 1954. The character & significance of sound production among fishes of the western North Atlantic. Bull. Bingham Oceanogr. Coll. 14(3):80-85.

Fish, M. and W. Mowbray. 1970. Sounds of Western North Atlantic Fishes. Baltimore:John Hopkins. p 156-158.

Fritzsche, R. 1978. Development of Fishes of the Mid-Atlantic Bight. Vol V. U.S. Fish Wildl. Ser. FWS/OBS 78/12. p 234-239.

Richards, S., J. Mann, and J. Walker. 1979. Comparison of spawning seasons, age, growth rates and food of two sympatric species of searobins, *Prionotus carolinus* and *P. evolans*, from Long Island Sound. Estuaries 2(4):255-268.

Russell, M., M. Grace and E. Gutherz. 1992. Field guide to searobins in the Western North Atlantic. NOAA Tech. Rept. NMSF 107. 26 pp.

Silver, W. and T. Finger. 1984. Electrophysiological examination of a non-olfactory, non-gustatory chemosense in the searobin, *Prionotus carolinus*. J. Comp. Physiol. 154:167- 174.

Northern star coral, *Astrangia poculata (=A. danae)*

Bachand, R. 1978. Cold water coral. Sea Frontiers. 24(5):282-284.

Barnes, R. 1987. Invertebrate Zoology. Philadelphia: Saunders. p 133-134.

Chapman, G. 1974. The skeletal system. In Coelenterate Biology. L. Muscatine and H. Lenhoff (eds.). New York: Academic Press. p 117.

Froehlich, A. 1980. Studies in the reproduction, nutrition and symbiosis with zooxanthellae of the temperate scleractinian coral *Astrangia dance*. Ph.D. dissertation. Univ. RI.

Jacques, Thierry. 1978. Metabolism and calcification of the temperate scleractinian coral *Astrangia danae*. Ph.D. dissertation. Univ. RI.

Muscatine, Leonard. 1974. Endosymbiosis of cnidarians and algae. In Coelenterate Biology, In L. Muscatine and H. Lenhoff (eds.) New York: Academic Press. p. 359-395.

Peters, E. et al. 1988. Nomenclature and biology of *Astrangia poculata*. Proc. Biol. Soc. Wash. 101(2):234-250.

Yonge, C. 1963. The biology of coral reefs. In Advances in Marine Biology, Vol 1. F.S. Russell (ed.) New York:Academic Press. p 209-260.

Oyster toadfish, *Opsanus tau*

Bachand, R. 1981. The oyster toadfish: A voice in Long Island Sound. Sea Frontiers. 27(3):179-183.

Breeder, H. and W. Schroeder. 1953. Fishes of the Gulf of Maine. U. S. Fish Wildl. Ser. Fish Bull. 53:518-520.

Fine, M. 1975. Sexual dimorphism and growth rate of the swimbladder of the toadfish, *Opsanus tau*. Copcia. 1975(3):483-490.

Fine, M. 1978. Seasonal and geographic variation of the mating call of the oyster toadfish, *Opsanus tau*. Oecologia. 36:45-57.

Fish, J. 1972. The effect of sound playback on the toadfish. In Behavior of Marine Animals, Vol. 2. H. Winn and B. Olla (eds.). New York: Plenum Press. p 386-434.

Fish, M. 1954. The character & significance of sound production among fishes of the western North Atlantic. Bull. Bingham Oceanogr. Coll. 14:80-85.

Fish, M. P. and W. Mowbray. 1970. Sounds of Western North Atlantic Fishes. Baltimore: John Hopkins Press. 205 pp.

Gray, G. and H. Winn. 1961. Reproductive ecology and sound production of the toadfish, *Opsanus tau*. Ecology. 42(2):274-282.

Grudger, E. 1910. Habits and life history of the toadfish, *Opsanus tau*. Fish. Bull. 28:1071-1109.

Isaacson, P. 1964. Summer movement of the toadfish, *Opsanus tau*. Ecology. 45(3):655-656.

Martin, F. and G. Drewry. 1978. Development of Fishes of the Mid-Atlantic Bight. Vol VI. U.S. Fish and Wildl. Ser. FWS/OBS-78-12. p 342-351.

Wilson, C. J. Dean and R. Radtke. 1982. Age, growth and feeding habits of the oyster toadfish, *Opsanus tau*. J. Exp. Mar. Biol. Ecol. 62:251-259.

Winn, H. 1972. Acoustic discrimination by the toadfish with comments on signal system. In Behavior of Marine Animals, Vol 2. H. Winn and B. Olla (eds.). New York: Plenum Press. p 361-385.

Portly spider crab, *Libinia emarginata*

Auster, P. and R. DeGoursey. 1983. Mass aggregation of Jonah and rock crabs. Bull. Am. Littoral Soc. 14(2):24-25.

Gutsell, J.S. 1928. The spider crab, *Libinia dubia*, and the jellyfish, *Stomolophus meleagris*, found in association at Beaufort, NC. Ecology 9(3):358-359.

Jackowski, R. 1963. Observations of the moon jelly, *Aurelia aurita*, and the spider crab, *Libinia dubia*. Chesapeake Sci. 4(4):195.

DeGoursey, R. and Stewart, L. 1985. Spider crab podding behavior and mass molting. Bull. Am. Littoral Soc. 15(2):12-16.

Hinsch, G.W. 1968. Reproductive behavior in the spider crab, *Libinia emarginata*. Biol. Bull. 135(2):273-278.

Jachowski, R. 1963. Observations on the moon jelly, *Aurelia aurita*, and the spider crab, *Libinia dubia*. Chesapeake Sci. 4(4):195.

Rice, A. 1988. The megalopa stage of majid crabs, with a review of spider crab relationships based on larval characters. In Aspects of Decapod Crustacean Biology. A. Finchman and P. Rainbow, (eds.). Zool. Symposium, No 59. 27-46.

Sandifer, P and W. Van Engel. 1971. Larval development of the spider crab, *Libinia dubia*, reared in laboratory culture. Chesapeake Sci. 12(1):18-25.

Shanks, A. and W. Graham. 1988. Chemical defense in a scyphomedusa. Mar. Ecol. Prog. Ser. 45:81-86.

Wicksten, M. 1980. Decorator crabs. Sci. Am. 242(2):146-154.

Williams, A. 1984. Shrimps, Lobsters and Crabs of the Atlantic Coast of the Eastern United States, Maine to Florida. Smithsonian Press. Washington, D.C. p 316-320.

Winget, R., D. Maurer, and H. Seymour. 1974. Occurrence, size composition and sex ratio of the rock crab, *Cancer irroratus* and the spider crab, *Libinia emarginata* in Delaware Bay. Natural History. 8(2):199-205.

Salt marsh snail (=Eastern melampus), *Melampus bidentatus*

Apley, M. 1970. Field studies on life history, gonadal cycle and reproductive periodicity in *Melampus bidentatus*. Malacologia, 10(2):381-397.

Daiber, F. 1982. Animals of the Tidal Marsh. New York: Van Norstrand Reinhold. 422 pp.

Hausman, S. 1932. A contribution to the ecology of the salt marsh snail, *Melampus bidentatus*. Am. Naturalist 66:541-545.

Holland, B. S. Loomis, and J. Gordon. 1991. Ice formation and freezing damage in the

174

foot muscle of the intertidal snail, *Melampus bidentatus*. Cryobiology. 28:491-498.

Holle, P. 1957. Life history of the salt-marsh snail, *Melampus bidentatus*. Nautilus 70:90-95.

Loomis, S. 1985. Seasonal changes in freezing tolerance of the intertidal pulmonate gastropod *Melampus bidentatus*. Canad. J. Zool. 63(9):2021-2025.

McMahon, R. and W. Russell-Hunter. 1981. Effects of physical variables and acclimation on survival and oxygen consumption in the high littoral salt marsh snail, *Melampus bidentatus*. Biol. Bull. 161:246-269.

Russell-Hunter, W., M. Apley, R. Hunter. 1972. Early life history of *Melampus* and the significance of semilunar synchrony. Biol. Bull. 143:623-656.

Sand fiddler crab, *Uca pugilator*

Bachand, R. 1979. The fascinating fiddler crab. Underwarer Naturalist. 11(4): 12-14.

Bright, D. and C. Hogue. 1972. A synopsis of burrowing land crabs of the world and list of their arthropod symbionts and burrow associates. Contr. Sci. 220:1-58.

Christy, J. 1980. The mating system of the sand fiddler, *Uca pugilator*. PhD dissertation, Cornell University, Ithaca, NY.

Crane, J. 1958. Aspects of social behavior in fiddler crabs, with special reference to *Uca maracoani*. Zoologica 43:113-130.

Crane, J. 1975. Fiddler crabs of the world. Ocypodidae: genus *Uca*. Princeton, NJ:Princeton Univ. Press. 736 pp.

Gray, E. 1942. Ecological and life history aspects of the red-jointed fiddler crab, *U. minax*, region of Solomon Island, MD. Chesapeake Biol. Lab. Publ. 51:3-20.

Powers, L. and D. Bliss. 1983. Terrestrial adaptations. In The Biology of Crustacea, Vol 8. F. Vernberg and W. Vernberg (eds.). New York:Academic Press. p 271-334.

Rebach, S. 1983. Orientation and migration in crustacea. In Studies in Adaptation, S. Rebach and D. Dunham (eds). New York:John Wiley & Sons. p 217-264.

Salmon, M. 1965. Waving display and sound production in *Uca minax* and *Uca pugnax*. Zoologica 50:123-150.

Salmon, M. 1968. Visual and acoustical signaling during courtship by fiddler crabs. Am. Zoologist 8:623-639.

Salmon, M. and J. Stout. 1962. Sexual discrimination and sound production in *Uca pugilator*. Zoologica 47:15-19.

Teal, J. 1958. Distribution of marsh crabs in Georgia salt marshes. Ecology 39(2): 185-193.

Zucker, H. 1977. Neighbor dislodgement and burrow-filling activities by male *Uca musica terpsichores:* A spacing mechanism. Mar. Biol. 41:282-286.

Sand tiger shark, *Carcharias taurus* and Sandbar shark, *Carcharhinus plumbeus*

Bigelow, H. and W. Schroeder. 1948. Sharks. In Fishes of the Western North Atlantic. Memoir No. 1, Sears Foundation of Marine Research. p. 59-375.

Casey, J. et al. 1991. Shark tagger summary. Newsletter coop. U.S. Dep. Commer. NOAA, NMSF, Narragansett, RI. 14 pp.

Casey, et al. 1992. Shark tagger summary. Newsletter coop. U.S. Dep. Commer. NOAA, NMSF, Narragansett, RI. 16 pp.

Casey, J. and L. Natanson. 1992. Revised estimates of age and growth of the sandbar shark, *Carcharhinus plumbeus*, from the Western North Atlantic. Can. J. Fish. Aquat. Sci. 49:1474-1477.

Gilbert, P. 1982. Patterns of shark reproduction. Oceanus. 24(4):23-29.

Gilmore, R. G., J. Dodrill, and P. Linley. 1983. Reproduction and embryonic development of the sand tiger shark, *Odontaspis taurus*. Fish. Bull. 81(2):201-225.

Gilmore, R. G. 1993. Reproductive biology of sharks. Env. Biol. Fish. 38:94-114.

Gilmore, G. 1994. Personal communications.

Gordon, I. 1993. Pre-copulatory behavior of captive sandtiger sharks, *Carcharias taurus*. Env. Biol. Fish. 38:159-164.

Greenwood, P. 1975. A History of Fishes. London:Enerst Benn Ltd.467 pp.

Gruber, S. and C. Manire. The only good shark is a dead shark? In Discovering Sharks S. Gruber (ed.). Sandy Hook, NJ: American Littoral Society. 121 pp.

Luer, C. and P. Gilbert, 1991. Elasmobranch fish. Oceanus 34(3):47-53.

Medved, R. and J. Marshall. 1983. Short-term movements of young sandbar sharks, *Carcharhinus plumbeus*. Bull. Mar. Sci. 33(1):87-93.

Musick, J. et al. 1993. Trends in shark abundance from 1974 to 1991 for the Chesapeake Bight region of the U.S. Mid-Atlantic coast. In Conservation Biology of Elasmobranchs, J. Branstetter (ed.). NOAA Tech. Rep. NMSF 115. p. 1-17.

Springer, S. 1948. Oviphagous embryos on the sand shark, *Carcharias taurus*. Copeia. 1948(3):153-175.

Springer, S. 1960. Natural History of the sandbar shark, *Eulamia milberti*. Fish. Bull. 61:1-38.

Springer, V. and J. Gold. 1989. Sharks in Question. Washington, DC: Smithsonian Inst. Press. 187 pp.

Stillwell, C. and N. Kohler. 1993. Food habits of the sandbar shark *Carcharhinus plumbeus* off the U. S. northeast coast, with estimates of daily ration. Fish. Bull. 91:138-150.

Sandworm, *Nereis virens.*

Bass, N. and A. Bradfield. 1972. The life cycle of the polychaete *Nereis virens*. J. Mar. Biol. Assn. U.K. 52:701-726

Dean, D. 1978. Migration of sandworms *Nereis virens* during winter nights. Mar. Biol. (Berl.) 45:165-173.

Snow, D. and J. Marsden. 1974. Life cycle, weight and possible age distribution in a population of *Nereis virens* from New Brunswick. J. Nat. Hist. 8:513-527.

Wilson, W. and R. Ruff. 1988. Species profiles: Sand worm and bloodworm. U.S. Fish Wildl. Serv. Biol Rep. 82(11.80) 23 pp.

Say mud crab, *Dyspanopeus (=Neopanope) sayi.*

Naylor, E. 1960. A North American xanthid crab new to Britain. Nature, 187:256-257.

Ryan, E. P. 1956. Observations on the life histories and distribution of the Xanthidae (mud crabs) of the Chesapeake Bay. Am. Midl. Nat. 56:138-162.

Swartz, R. 1976. Sex ratio as a function of size in the xanthid crab, *Neopanope sayi*. Am. Nat. 110(975):898-902.

Swartz, R. 1976. Agonistic and sexual behavior of the xanthid crab, *Neopanope sayi*. Chesapeake Sci. 17(1):24-34

Swartz, R. 1978. Reproductive and molt cycles in the xanthid crab, *Neopanope sayi*. Crustaceana 34(1):15-32.

Sevenspine bay shrimp (=sand shrimp), *Crangon septemspinosa.*

Embich, T. 1973. Ecology of the sand shrimp *Crangon septemspinosa* in Penobscot Bay, ME. M.S. thesis. Univ ME, Orono.59 p.

Haefner, P. 1971. Avoidance of anoxic conditions by the sand shrimp, *Cragon septemspinosa*. Chesapeake Sci. 12(1):50-51.

Haefner, P. 1972. The biology o the sand shrimp, *Crangon septemspinosa* at Lamoine, ME. J. Elisha Mitchell Soc. 88:36-42.

Haefner, P.A. 1979. Comparative review of the biology of North Atlantic caridean shrimps (*Crangon*) with emphasis on *Crangon septemspinosa*. Bull. Biol. Soc. Washington.3:1-40.

Herman, S. 1963. Vertical migration of the opossum shrimp, *Neomysis americana*. Limnol. Oceanogr. 8(3):228-238.

Modlin, R. 1976. Life history, ecology and population dynamics of *Crangon septemspinosa* in the Mystic River Estuary, CT. Ph.D. dissertation, Univ. CT, Storrs, CT. 91 pp.

Modlin, R. 1980. The life cycle and recruitment of the sand shrimp, *Crangon septemspinosa*, in the Mystic River estuary, CT. Estuaries, 3(1):1-10.

Needler, A. 1941. Larval stages of *Crago septemspinosa*. Trans. Roy Can. Inst. 23:193-199.

Normandeau Associates, Inc. 1979. New Haven Harbor Ecological Studies. Summary Report 1970-1977. United Illuminating Co., New Haven, CT.

Price, K. 1962. Biology of the sand shrimp, *Crangon septemspinosa*, in the shore zone of the Delaware Bay region. Chesapeake Sci. 3:244-255.

Richards, S. and G. Riley. 1967. The epifauna of Long Island Sound. Bull. Bingham Oceanogr. Coll. 19:89-135.

Welsh, B. 1970. Some aspects of the vertical migration and predatory behavior of the sand shrimp, *Crangon septemspinosa*. MS thesis, Univ. MD. 66 pp.

Wilcox, J. and H. Jeffries. 1973. Growth of the sand shrimp, *Crangon septemspinosa*, in Rhode Island. Chesapeake Sci. 14(3):201-205.

Wilcox, J. and H. Jeffries. 1974. Feeding habits of the sand shrimp, *Crangon septemspinosa*. Biol. Bull. 146:424-434.

Sheepshead minnow, *Cyprinodon variegatus*

Able, K. 1976. Cleaning behavior in the Cyprinodontid fishes: *Fundulus majalis*, *Cyprinodon variegatus* and *Lucania parva*. Chesapeake Sci. 17(1):35-39.

Abraham, B. 1985. Species profile: Mummichog and striped killifish. U.S. Fish Wildl. Serv. Biol. Bull. Rep. 82(11.40). 23 pp.

McCuthcheon, F. and A. McCuthcheon. 1964. Symbiotic behavior among fishes from temperate waters. Science. 145:948-949.

Raney, E. et al. 1953. Reproductive behavior of *Cyprinodon variegatus* in Florida. Zoologica. 38:97-104.

Tyler, A. 1963. A cleaning symbiosis between rainwater fish, *Lucania parva* and the stickleback, *Apeltes quadracus*. Chesapeake Sci. 4:105-108.

Spider crab (see Portly spider crab)

Striped bass, *Morone (=Roccus) saxatilis*

Bain, M. and J. Bain. 1982. Habitat suitability index models: Coastal stocks of striped bass. U. S. Fish Wildl. Serv. FWS/OBS-82/10.1. 29 pp.

Bigelow, H and W. Schroeder. 1953. Fishes of the Gulf of Maine. Fish Wildl. Serv. Bull. 53:389-404.

Fay, C., R. Neves, and G. Pardue. 1983. Species profiles: Mid-Atlantic striped bass. U. S. Fish Wildl. Serv. FWS/OBS 82/11.8. 36 pp.

Hardy, J. 1978. Development of fishes of the Mid-Atlantic Bight. Vol. lll U.S. Fish Wildl. Serv. FWS/OBS-78/12. p 86-105.

Hill, J., J. Evans, and M. Van Den Avyle. 1989. Species profile: (South Atlantic) striped bass. U.S. Fish Wildl. Serv. Biol. Rep. 82(11.118) 35 pp.

Kohlenstein, L. 1981. On the proportion of the Chesapeake Bay stock of striped bass that migrates into the coastal fishery. Trans. Am. Fish. Soc. 110:168-179.

McLaren, J. et al. 1981. Movements of Hudson River striped bass. Trans. Am. Fish. Soc. 110:158-167.

Raney, E. 1952. The life history of the striped bass. Bingham Oceanogr. Collect. Yale Univ. Bull. 14:5-97.

Rathjen, W. and L. Miller. 1957. Aspects of the early life history of the striped bass, *Roccus saxatilis*, in the Hudson River. NY Fish Game J. 4(1):43-60.

177

Setzler, E. et al. 1980. Synopsis of biological data on striped bass. Nat. Mar. Fish. Serv. FAO Synopsis No 121. 69 pp.

Striped killifish, *Fundulus majalis*
Abraham, B. 1985. Species profile: Mummichog and striped killifish. U.S. Fish Wildl. Serv. Biol. Rep. 82(11.40). 23 pp.

Bigelow, H. and W. Schroeder. 1953. Fishes of the Gulf of Maine. U. S. Fish Wildl. Ser. Fish Bull. 53:164-165.

Mast, S. 1915. The behavior of *Fundulus*, with especial reference to overland escape from tidepools and locomotion over land. J. Animal Behavior 5(5):341-350.

Newman, H. 1907. Spawning behavior and sexual dimorphism in *Fundulus heteroclitus* and allied fish. Biol. Bull. 12:314-345.

Newman, H. 1908. A significant case of hermaphroditism in fish. Biol. Bull. 15(5):207-214.

Striped searobin, see Northern searobin

Summer flounder, *Paralichthys dentatus*
Able, K. et al. 1989. Patterns of summer flounder, *Paralichthys dentatus* early life history in the Mid-Atlantic Bight and New Jersey estuaries. Fish. Bull. 88:1-12.

Bigelow, H. and W. Schroeder. 1953. Fishes of the Gulf of Maine. Fish. Bull. 53:267-270.

Martin, F. 1978. Development of Fishes of the Mid-Atlantic Bight, Vol. VI. U. S. Fish and Wildl. Serv. FWS/OBS-78-12. p 157-163.

Morse, W. 1981. Reproduction of the summer flounder *Paralichthys dentatus*. J. Fish Biol. 19:189-203.

Powell, A. 1974. Biology of the summer flounder, *Paralichthys dentatus*, in Pamlico Sound. M.S. thesis. Univ. North Carolina, Chapel Hill. 145 pp.

Powell, A. and F. Schwartz. 1977. Distribution of paralichtid flounders in North Carolina estuaries. Chesapeake Sci. 18:334-339.

Rogers, S. and M. Van Den Avyle. 1983. Species profile: Summer flounder. U.S. Fish and Wildl. Serv. FWS/OBS-82/11.15 14 pp.

Rountree, R. and K. Able. 1992. Foraging habits, growth and temporal patterns of salt-marsh creek habitat use by young-of-the-year summer flounder in New Jersey. Trans. Am. Fish. Soc. 121:765-776.

Smith, R. and F. Daiber. 1977. Biology of the summer flounder, *Paralichthys dentatus*, in Delaware Bay. Fish. Bull. 75(4):823-830.

Szedlmayer, S., K. Able, and R. Rountree. 1992. Growth and temperature-induced mortality of young-of-the-year summer flounder, *Paralichthys dentatus*, in southern New Jersey. Copeia. 1992(1):120-128.

Weinstein, M. et al. 1980. Retention of three taxa of post-larval fishes in an intensively flushed tidal estuary, Cape Fear, NC. Fish. Bull. 78:419-436.

Tautog(=blackfish), *Tautoga onitis*
Bejda, A. Personal communications, 1993.

Bigelow, H. and W. Schroeder. 1953. Fishes of the Gulf of Maine. Fish. Bull. 53:478-483.

Connecticut DEP. 1989. Principal fisheries of Long Island Sound. 47 pp.

Cooper, R. 1966. Migration and population estimation of the tautog, *Tautoga onitis* from Rhode Island. Trans. Am. Fish. Soc. 95:239-247.

Cooper, R. 1967. Age and growth of the tautog, *Tautoga onitis* from Rhode Island. Trans. Am. Fish. Soc. 96:134-142.

Eklund, A. and T. Targett. 1990. Reproductive seasonality of fishes inhabiting hard bottom areas in the Middle Atlantic Bight. Copeia 1990(4):1180-1184.

Fritzsche, R. 1978. Development of the Fishes of the Mid-Atlantic Bight. Vol V. U. S. Department of the Interior, FWS/OBS 78-/12. p 29-35.

Grover, J. 1982. The comparative feeding ecology of five inshore, marine fishes off Long Island, NY. Ph.D. dissertation, Rutgers Univ. New Brunswick, 197 p.

Olla, B., A. Bejda, and A. Martin. 1974. Daily activity, movements, feeding and seasonal occurrences of the tautog, Tautoga onitis. Fish. Bull. 77:255-261.

Olla, B. and C. Samet. 1977. Courtship and spawning behavior of the tautog, Tautoga onitis under laboratory conditions. Fish. Bull. 75(3):585-599.

Olla, B., A. Studholme, A. Bejda, and A. Martin. 1978. Effect of temperature on activity and social behavior of the adult tautog Tautoga onitis under laboratory conditions. Mar. Biol. (Berl.) 45:369-378.

Olla, B., A. Bejda and A. Martin. 1979. Seasonal dispersal and habitat selection of cunner, Tautogolabrus adspersus, and young tautog, Tautoga onitis, in Fire Island Inlet, LI, NY. Fish. Bull. 77(1):255-261.

Olla, B., A. Studholme, A. Bejda, and C. Samet. 1980. Role of temperature in triggering migratory behavior of the adult tautog Tautoga onitis under laboratory conditions. Mar. Biol. (Berl.) 59:23-30.

Olla, B., C. Samet and A. Studholme. 1981. Correlates between number of mates, shelter availability and reproductive behavior in the tautog, Tautoga onitis. Mar. Biol. (Berl.) 62:239-248.

Sogard, S. and K. Able. 1991. A comparison of eelgrass, sea lettuce macroalgae and marsh creeks as habitats for epibenthic fishes and decapods. Estuarine Coastal Shelf Sci. 33:501-519.

Sogard, S., K. Able, and M. Fahay, 1992. Early life history of the tautog Tautoga onitis in the Mid-Atlantic Bight. Fish. Bull. 90(3):529-539.

Wheatland, S. 1956. Oceanography of Long Island Sound, 1952-1954. VII. Pelagic fish eggs and larvae. Bull. Bingham oceanogr. Coll. 15:234-314.

Winter flounder, Pleuronectes (=Pseudopleuronectes) americanus

Bigelow, H.B. and W. C. Schroeder. Fishes of the Gulf of Maine. Fish. Bull. 53:276-283.

Breder, C.M. 1922. Description of spawning habits of Pseudopleuronectes americanus in captivity. Copeia 102:3-4.

Breder, C.M. 1924. Some embryonic and larval stages of the winter flounder. Bull. Bur. Fish. 38:311-315.

Crawford, R. 1990. Winter flounder in Rhode Island coastal ponds. Rhode Island Sea Grant, National Sea Grant Depository Publ #RIU-G-90-001. 23 pp.

Duman, J. and A. deVries. 1974. Freezing resistance in winter flounder, Pseudopleuronectes americanus. Nature (Lond.) 247:247-248.

Grimes, B.et al. 1989. Species profiles: Summer and winter flounder. U. S. Fish. Wildl. Serv. Biol. Rep. 82(11.112). 18 pp.

Howe, A., P. Coates, and D. Pierce. 1976. Winter flounder estuarine year-class abundance, mortality, and recruitment. Trans. Am. Fish. Soc. 105(6):647-657.

Klein-MacPhee, G. 1978. Synopsis of biological data for the winter flounder, Pseudopleuronectes americanus. US NOAA NMSF Tech. Report Circ. 414. 43 pp.

McCraken, F. 1963. Seasonal movements of the winter flounder, Pseudopleuronectes americanus on the Atlantic coast. J. Fish. Res. Bd. Canada 20(2):551-586.

Martin, F.D. and G. E. Drewry. 1978. Development of fishes of the mid-Atlantic bight, Vol VI. FWS/OBS-78/12. p 197-200.

Olla, B., R. Wicklund and S. Wilk. 1969. Behavior of the winter flounder in a natural habitat. Trans. Am. Fish. Soc. 98:717-720.

Phelan, B. 1992. Winter flounder movements in the inner New York Bight. Trans. Am. Fish. Soc. 121:777-784.

179

Pierce, D. and A. Howe. 1977. A further study on winter flounder group identification off Massachusetts. Trans. Am. Fish. Soc. 106(3):131-139.

Saila, S. 1961. A study of winter flounder movements. Limn. Oceanogr. 6:292-298.

Scott, W. 1929. A note on the effect of temperature and salinity on the hatching of eggs of the winter flounder. Contrib. Canad. Biol. Fish. Stat. Can. N.S. 4:139-141.

Sullivan, W.E. 1914. A description of the young stages of the winter flounder, *Pseudopleuronectes americanus*. Trans. Am. Fish. Soc. 44:125-136.

Wells, B., D. Steele and A. Tyler. 1973. Intertidal feeding of winter flounder, *Pseudopleuronectes americanus*, in the Bay of Fundy. J. Fish. Res. Board Can. 30:1374-1378.

Index